# ASSESSMENT HANDBOOK

# Everyday
# Mathematics®

The University of Chicago School Mathematics Project

Mc
Graw
Hill
Education

# The University of Chicago School Mathematics Project

Max Bell, Director, *Everyday Mathematics* First Edition; James McBride, Director, *Everyday Mathematics* Second Edition; Andy Isaacs, Director, *Everyday Mathematics* Third, CCSS, and Fourth Editions; Amy Dillard, Associate Director, *Everyday Mathematics* Third Edition; Rachel Malpass McCall, Associate Director, *Everyday Mathematics* CCSS and Fourth Editions; Mary Ellen Dairyko, Associate Director, *Everyday Mathematics* Fourth Edition

## Authors
Jean Bell, Max Bell, John Bretzlauf, Mary Ellen Dairyko, Amy Dillard, Robert Hartfield, Andy Isaacs, Kathleen Pitvorec, James McBride, Peter Saecker

## Fourth Edition Grade 3 Team Leader
Mary Ellen Dairyko

## Writers
Lisa J. Bernstein, Camille Bourisaw, Julie Jacobi, Gina Garza-Kling, Cheryl G. Moran, Amanda Louise Ruch, Dolores Strom

## Open Response Team
Catherine R. Kelso, Leader; Amanda Louise Ruch, Andy Carter

## Differentiation Team
Ava Belisle-Chatterjee, Leader; Martin Gartzman, Barbara Molina, Anne Sommers

## Digital Development Team
Carla Agard-Strickland, Leader; John Benson, Gregory Berns-Leone, Juan Camilo Acevedo

## Virtual Learning Community
Meg Schleppenbach Bates, Cheryl G. Moran, Margaret Sharkey

## Technical Art
Diana Barrie, Senior Artist; Cherry Inthalangsy

## UCSMP Editorial
Lila K. S. Goldstein, Senior Editor; Kristen Pasmore, Molly Potnick, Rachel Jacobs

## Field Test Coordination
Denise A. Porter

## Field Test Teachers
Eric Bachmann, Lisa Bernstein, Rosemary Brockman, Nina Fontana, Erin Gilmore, Monica Geurin, Meaghan Gorzenski, Deena Heller, Lori Howell, Amy Jacobs, Beth Langlois, Sarah Nowak, Lisa Ringgold, Andrea Simari, Renee Simon, Lisa Winters, Kristi Zondervan

## Digital Field Test Teachers
Colleen Girard, Michelle Kutanovski, Gina Cipriani, Retonyar Ringold, Catherine Rollings, Julia Schacht, Christine Molina-Rebecca, Monica Diaz de Leon, Tiffany Barnes, Andrea Bonanno-Lersch, Debra Fields, Kellie Johnson, Elyse D'Andrea, Katie Fielden, Jamie Henry, Jill Parisi, Lauren Wolkhamer, Kenecia Moore, Julie Spaite, Sue White, Damaris Miles, Kelly Fitzgerald

## Contributors
John Benson, Jeanne Mills DiDomenico, James Flanders, Lila K. S. Goldstein, Funda Gonulates, Allison M. Greer, Catherine R. Kelso, Lorraine Males, Carole Skalinder, Jack Smith, Stephanie Whitney, Penny Williams, Judith S. Zawojewski

## Center for Elementary Mathematics and Science Education Administration
Martin Gartzman, Executive Director; Meri B. Fohran, Jose J. Fragoso, Jr., Regina Littleton, Laurie K. Thrasher

## External Reviewers

The *Everyday Mathematics* authors gratefully acknowledge the work of the many scholars and teachers who reviewed plans for this edition. All decisions regarding the content and pedagogy of *Everyday Mathematics* were made by the authors and do not necessarily reflect the views of those listed below.

Elizabeth Babcock, California Academy of Sciences; Arthur J. Baroody, University of Illinois at Urbana-Champaign and University of Denver; Dawn Berk, University of Delaware; Diane J. Briars, Pittsburgh, Pennsylvania; Kathryn B. Chval, University of Missouri–Columbia; Kathleen Cramer, University of Minnesota; Ethan Danahy, Tufts University; Tom de Boor, Grunwald Associates; Louis V. DiBello, University of Illinois at Chicago; Corey Drake, Michigan State University; David Foster, Silicon Valley Mathematics Initiative; Funda Gönülateş, Michigan State University; M. Kathleen Heid, Pennsylvania State University; Natalie Jakucyn, Glenbrook South High School, Glenview, IL; Richard G. Kron, University of Chicago; Richard Lehrer, Vanderbilt University; Susan C. Levine, University of Chicago; Lorraine M. Males, University of Nebraska-Lincoln; Dr. George Mehler, Temple University and Central Bucks School District, Pennsylvania; Kenny Huy Nguyen, North Carolina State University; Mark Oreglia, University of Chicago; Sandra Overcash, Virginia Beach City Public Schools, Virginia; Raedy M. Ping, University of Chicago; Kevin L. Polk, Aveniros LLC; Sarah R. Powell, University of Texas at Austin; Janine T. Remillard, University of Pennsylvania; John P. Smith III, Michigan State University; Mary Kay Stein, University of Pittsburgh; Dale Truding, Arlington Heights District 25, Arlington Heights, Illinois; Judith S. Zawojewski, Illinois Institute of Technology

## Note
Many people have contributed to the creation of *Everyday Mathematics*. Visit http://everydaymath.uchicago.edu/authors/ for biographical sketches of *Everyday Mathematics* 4 staff and copyright pages from earlier editions.

## www.everydaymath.com

Send all inquiries to:
McGraw-Hill Education
8787 Orion Place
Columbus, OH 43240

ISBN: 978-0-02-130757-9
MHID: 0-02-130757-1

Printed in the United States of America.

1 2 3 4 5 6 7 8 9 RHR 20 19 18 17 16 15

# Contents

# Assessment in *Everyday Mathematics*®

Assessment in *Everyday Mathematics*:

- addresses the full range of content and practices in the Common Core State Standards (CCSS).

- consists of tasks that are worthwhile learning experiences.

- is manageable for teachers.

- informs instruction by providing actionable information about children's progress.

- provides information for grading.

- clarifies the *Everyday Mathematics* spiral and helps teachers decide when to intervene and when "watchful waiting" is appropriate.

- serves basic Tier 1 and Tier 2 Response to Intervention (RtI) functions.

- provides information that will complement data from standards-based assessments, including those from PARCC and SBAC.

**Go Online** for more information in the Assessment section of the *Implementation Guide*.

## Assessment of Content and Practices

*Everyday Mathematics* integrates instruction and assessment of mathematical practices with instruction and assessment of grade-level content. The mathematical practices are not to be separated from the content; they are mathematical habits of mind children should develop as they learn mathematical content.

However, the content and practice standards in the CCSS differ in important respects. The content standards describe specific goals that are organized by mathematical domain and differ from grade to grade. The practice standards describe general, cross-grade goals that are related to processes such as problem solving, reasoning, and modeling. Many tasks in *Everyday Mathematics* provide opportunities to assess both content and practice standards. However, due to the differing nature of these standards, *Everyday Mathematics* assesses and tracks progress on them in different ways.

### Assessing the Content Standards

Each grade's Common Core content standards are unpacked into 50 to 80 *Everyday Mathematics* Goals for Mathematical Content (GMC). The standards and the corresponding GMCs are listed in the back of the *Teacher's Lesson Guide*. Instructional activities and assessment items are linked to one or more of the GMCs, which provide more targeted information for assessment and differentiation.

For each task that assesses a content standard, *Everyday Mathematics* provides guidance on what constitutes "meeting expectations" for that standard at that point in the year. Individual Profiles of Progress, Class Checklists, and the assessment and reporting tools help teachers monitor children's progress using this framework.

## Assessing the Practice Standards

Since the Standards for Mathematical Practice in the CCSS are broadly written for Grades K to 12, *Everyday Mathematics* includes Goals for Mathematical Practice (GMP) that unpack the Standards for Mathematical Practice for elementary school teachers and children. These *Everyday Mathematics* GMPs can be useful for assessing the CCSS mathematical practices because they highlight different aspects of each practice.

Tracking progress on the mathematical practices requires a more qualitative approach. Assessment opportunities include writing/reasoning prompts, open response problems, and observations of children in the course of daily work. Tools for assessing the practices include checklists and task-specific rubrics for open response problems.

**Go Online** for more information about the GMCs and GMPs in the *Everyday Mathematics* and the Common Core State Standards section of the *Implementation Guide*.

# Assessment Opportunities

*Everyday Mathematics* offers a variety of opportunities for ongoing and periodic assessment of content and practices.

## Assessment Check-Ins

Assessment Check-Ins are lesson-embedded opportunities to assess the focus content and practices of the lesson. They appear in regular lessons and Open Response and Reengagement lessons. Each Assessment Check-In provides information on expectations for particular standards at that point in the curriculum. The results can be used to inform instruction and, often, for grading.

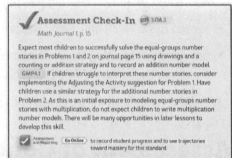

## Preview Math Boxes

One pair of Preview Math Boxes appears near the end of each unit. The Preview Math Boxes can be used to gauge children's readiness for the upcoming unit so that teachers can better plan instruction and choose appropriate differentiation activities.

## Writing/Reasoning Prompts

Many Math Boxes have Writing/Reasoning prompts that encourage children to communicate their understanding of concepts and skills and their strategies for solving problems. Writing/Reasoning prompts provide valuable opportunities for assessing the mathematical practices.

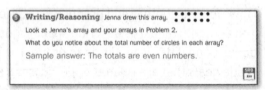

## Progress Checks

The last lesson in each unit is a Progress Check, a two-day lesson that provides the following assessment opportunities:

- **Self Assessment** Every unit includes this opportunity for children to consider how well they are doing on the focus content of the unit.

- **Unit Assessment** Every unit includes this assessment of the content and practices that were the focus of the unit. All items are appropriate for grading because they match the expectations for the standards they assess up to that point in the year.

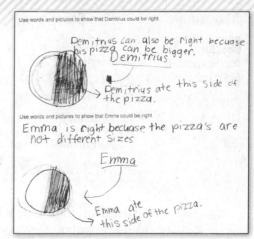

- **Open Response Assessment** Odd-numbered units include this opportunity for children to think creatively about a problem. It addresses one or more content standards and one Goal for Mathematical Practice that can be evaluated using a task-specific rubric.

- **Cumulative Assessment** Even-numbered units include this assessment of standards from prior units. All items are appropriate for grading because they match the expectations for progress on those standards at that point in the grade.

- **Challenge** Each Progress Check lesson includes one or more challenge problems related to important ideas from the unit.

## Interim Assessments

These assessments are administered at the beginning, middle, and end of the school year.

- **Beginning-of-Year Assessment** This assessment provides information about children's knowledge and skills related to the content in the first two or three units based on grade-level expectations from the prior grade. It can be useful for RtI screening.

- **Mid-Year and End-of-Year Assessments** These assessments offer teachers snapshots of children's performance on representative samples of standards covered to date. All items are appropriate for grading because they match the expectations for the standards assessed up to that point in the year.

## Summaries of Assessment Opportunities

Every Unit Organizer includes information about assessment opportunities in the unit. Every Progress Check has tables that summarize the content and practice standards assessed in the lesson.

## Other Assessment Opportunities

Almost any task in *Everyday Mathematics* can provide information that could be useful for assessment. Assessment tools in *Everyday Mathematics* can accommodate data from sources other than those listed above, and teachers should use their judgment about expanding the range of data they gather and use to inform their instruction and assign grades.

# Assessment Tools

*Everyday Mathematics* provides a variety of tools for collecting, storing, analyzing, reporting, and using assessment data.

## Rubrics for Evaluating Mathematical Practices

The Open Response and Reengagement lessons and the Open Response Assessments are opportunities to assess children's progress on the CCSS Standards for Mathematical Practice. Each of these includes a task-specific rubric that can be used to evaluate children's work for a specific *Everyday Mathematics* Goal for Mathematical Practice. See the sample rubric on the following page.

## Sample Rubric

| Goal for Mathematical Practice | Not Meeting Expectations | Partially Meeting Expectations | Meeting Expectations | Exceeding Expectations |
|---|---|---|---|---|
| **GMP3.1**<br><br>Make mathematical conjectures and arguments. | Argues that Demitrius could be correct or that Emma could be correct, but not both, or does not make an argument. | Argues that Demitrius could be correct and that Emma could be correct, but the words or drawings used to support these arguments are incomplete or unclear. | Uses either words or drawings to clearly argue that:<br>• Demitrius could be correct if his pizza had been bigger than Emma's, and<br>• Emma could be correct if they started with the same size pizza. | Meets expectations and clearly makes both arguments using both words and drawings. |

Generic rubrics for the Goals for Mathematical Practice that you can complete and use to evaluate children's responses to Writing/Reasoning prompts and certain items in the Assessment Check-Ins, Open Response and Reengagement lessons, and Progress Checks can be found in the appendix of this handbook.

**Go Online** for information on how to use the rubrics in the Assessment section of the *Implementation Guide*.

## Individual Profiles of Progress and Class Checklists

Individual Profiles of Progress (IPPs) combine data from various sources for individual children. Class Checklists facilitate collecting and recording data for an entire class. Blank masters of these forms are provided in this handbook.

**Go Online** for unit-specific versions of the IPPs and Class Checklists that you can download and print or complete digitally.

Individual Profile of Progress

Class Checklist

## Assessment Masters and Answer Keys

This handbook includes masters for the Progress Check assessments and the interim assessments. Answers for the Progress Check assessments are provided at point of use in the *Teacher's Lesson Guide*. Answers for the interim assessments are provided on pages 165–169 of this handbook.

**Go Online** for answer keys that you can view at full size.

## Assessment and Reporting Tools

Digital tools available to teachers through McGraw-Hill's ConnectEd platform centralize evaluation, reporting, and targeted differentiation. Teachers are able to evaluate children's work, whether completed digitally or in print, and generate reports on children's progress based on standards covered in lessons and units. The system can track children's performance and provide teachers information and access to materials at point of use to support differentiation decisions.

# Unit 1 Self Assessment

Put a check in the box that tells how you do each skill.

| Skills | I can do this on my own and explain how to do this. | I can do this on my own. | I can do this if I get help or look at an example. |
|---|---|---|---|
| ① Find the difference between two numbers on a number grid. **MJ1 3** | | | |
| ② Look for information in my *Student Reference Book.* **MJ1 4** | | | |
| ③ Round numbers. **MJ1 6** | | | |
| ④ Tell time. **MJ1 8** | | | |
| ⑤ Make a bar graph. **MJ1 13** | | | |
| ⑥ Solve multiplication number stories. **MJ1 15** | | | |

# Unit 1 Assessment

**①** Use the number grid.

| 91 | 92 | 93 | 94 | 95 | 96 | 97 | 98 | 99 | 100 |
|----|----|----|----|----|----|----|----|----|-----|
| 101 | 102 | 103 | 104 | 105 | 106 | 107 | 108 | 109 | 110 |
| 111 | 112 | 113 | 114 | 115 | 116 | 117 | 118 | 119 | 120 |
| 121 | 122 | 123 | 124 | 125 | 126 | 127 | 128 | 129 | 130 |

**a.** The difference between 95 and 127 is _____.

**b.** The difference between 97 and 122 is _____.

**c.** Explain how you used the number grid to solve Problem 1b.

_____

_____

_____

**②** Write the time shown on each clock.
You may use your toolkit clock to help you.

**a.**

**b.**

_____  _____

# Unit 1 Assessment (continued)

③ **a.** Use the tally chart to complete the bar graph.

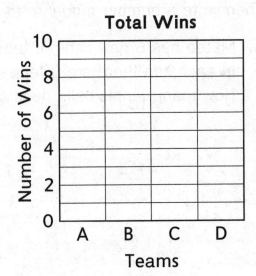

| Total Wins | |
|---|---|
| **Teams** | **Number of Wins** |
| Team A | ~~HHH~~ //// |
| Team B | ~~HHH~~ / |
| Team C | ~~HHH~~ |
| Team D | ~~HHH~~ ~~HHH~~ |

**Total Wins**

Use the data in the bar graph to answer the questions below.

**b.** How many wins did the four teams have in all? _____

**c.** How many fewer wins did Team C have than Team D? _____

④ Solve each problem.

**a.** $2 \times 5 =$ _____       **b.** $2 \times 8 =$ _____

**c.** $5 \times 3 =$ _____       **d.** $4 \times 5 =$ _____

**e.** $10 \times 2 =$ _____       **f.** $3 \times 10 =$ _____

**g.** How did you solve $4 \times 5$?

_____

_____

_____

# Unit 1 Assessment (continued)

(5) For each number story, draw a sketch and write the answer.
Then write a number model to fit the story.

**a.** Mateo has 6 new cans of tennis balls.
In each can there are 3 tennis balls.
How many tennis balls does Mateo have in all?

He has _____ tennis balls.

Number model: _____

**b.** Anne sketches 5 rows of flowers on her page with 6 flowers
in each row. How many flowers does she sketch in all?

She sketches _____ flowers.

Number model: _____

## Unit 1 Assessment (continued)

**6** Angela starts dance practice at 3:05 P.M. and finishes at 3:55 P.M. She drew an open number line and used it to find the length of her practice.

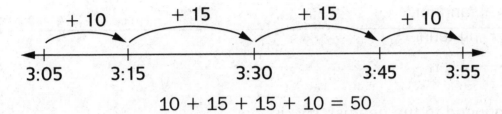

$$10 + 15 + 15 + 10 = 50$$

Explain Angela's work. _____

_____

_____

_____

_____

How long is Angela's dance practice? _____ minutes long

**7** Round each number to the nearest 10. You may use open number lines to help.

**a.** 59 rounded to the nearest 10 is _____.

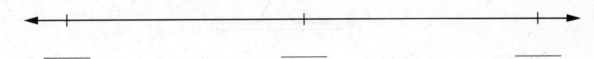

**b.** 73 rounded to the nearest 10 is _____.

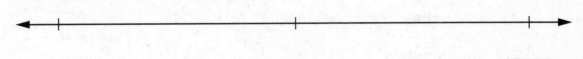

# Unit 1 Assessment (continued)

⑧ Round each number to the nearest 100.
You may use open number lines to help.

**a.** 423 rounded to the nearest 100 is _____.

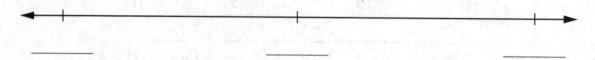

_____        _____        _____

**b.** 379 rounded to the nearest 100 is _____.

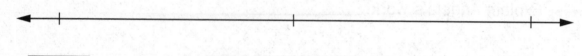

_____        _____        _____

⑨ Mary used a pan balance and masses to measure the
mass of a tennis ball. She put the tennis ball in one pan
and a 50-gram mass in the other pan. Then she added
one 5-gram mass and two 1-gram masses to balance
the pans. What is the mass of the tennis ball?

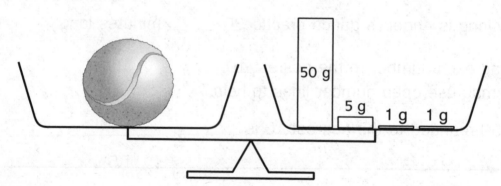

Answer: _____ grams

How did you figure out your answer? _____

_____

_____

# Unit 1 Challenge

① Marsha counts 20 blocks and arranges them in different arrays.

    **a.** Sketch all the possible arrays Marsha could make with the blocks.

    **b.** Write multiplication number models for each of the arrays.

    **c.** Could Marsha make an array that has 3 rows? _____

        Explain. _____

        _____

        _____

② Don and Molly played *Number-Grid Difference.*
    The object of the game is to have the lower sum of 5 scores.

    Don picked 3 and 5 and made the number 35.

    Molly picked 8 and 5. What number should Molly make? _____

    Explain your answer. _____

    _____

    _____

# Unit 1 Challenge (continued)

③ Solve. You may use your toolkit clock or an open number line to help you. Show your work.

Evan starts basketball camp at 9:15 A.M.
He finishes at 3:45 P.M.

How many hours and minutes does Evan spend at camp?

Evan spends _____ hours and _____ minutes at camp.

④ Manuel is working on his 10s and 5s facts.
He knows most of his 10s facts, but he has trouble with his 5s facts.
You can help him.

**a.** Solve.

$6 \times 10 =$ _____

$6 \times 10$ means 6 equal groups of 10.

**b.** Explain how Manuel can use his answer to $6 \times 10$ to figure out what $6 \times 5$ would be.

_____

_____

**c.** Explain another way that Manuel could solve $6 \times 5$.

_____

_____

# Unit 1 Open Response Assessment
## Getting to School

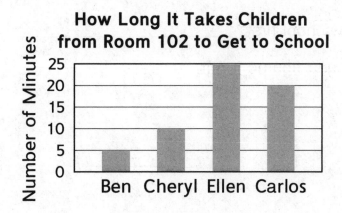

**How Long It Takes Children from Room 102 to Get to School**

Number of Minutes

Ben  Cheryl  Ellen  Carlos

(1) Look carefully at the title, labels, and bars on the graph. Write at least 5 things you know from the graph.

# Unit 1 Open Response Assessment (continued)

② Carlos leaves for school at 8:00 A.M. Cheryl leaves 5 minutes later.

**a.** Who gets to school first? _____

**b.** Explain how you figured it out.

_____

_____

_____

_____

_____

# Unit 2 Self Assessment

Put a check in the box that tells how you do each skill.

| Skills | I can do this on my own and explain how to do this. | I can do this on my own. | I can do this if I get help or look at an example. |
|---|---|---|---|
| ① Solve extended facts. **MJ1** 32 | | | |
| ② Solve number stories by adding or subtracting. **MJ1** 35–36 38–39 | | | |
| ③ Check whether my answer makes sense. **MJ1** 38–39 | | | |
| ④ Solve equal-groups and array number stories. **MJ1** 46 58 | | | |
| ⑤ Solve division number stories. **MJ1** 52 | | | |
| ⑥ Solve Frames-and-Arrows problems. **MJ1** 57 | | | |

# Unit 2 Assessment

Fill in the unit box. Then solve.

| Unit |
| --- |
|  |

**1**   **a.** $3 + \underline{\hspace{3em}} = 12$

     **b.** $30 + \underline{\hspace{3em}} = 120$

     **c.** $300 + \underline{\hspace{3em}} = 1{,}200$

**2**   **a.** $17 - 8 = \underline{\hspace{3em}}$

     **b.** $27 - 8 = \underline{\hspace{3em}}$

     **c.** $57 - 8 = \underline{\hspace{3em}}$

Fill in the missing rule or numbers.

**3**

| Rule |
| --- |
| $-10$ |

74 → ___ → 54 → ___ → ___ → 24

**4**

| Rule |
| --- |
|  |

2 → 4 → 8 → ___ → ___ → 64

# Unit 2 Assessment (continued)

For each number story, write a number model with a ?.
Then solve the number story.
You may draw diagrams, like those below, or pictures to help.

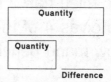

(5) Maria swam a total of 56 minutes over the weekend. She swam for
20 minutes on Saturday. How many minutes did she swim on Sunday?

_____
(number model with ?)

Answer: _____
(unit)

How do you know your answer makes sense?

_____

_____

(6) One python clutch has 31 eggs. Another python clutch has 19 eggs.
How many more eggs are in the first clutch?

_____
(number model with ?)

Answer: _____
(unit)

How do you know your answer makes sense?

_____

_____

# Unit 2 Assessment (continued)

⑦ Jeremiah read the number story below. Then he drew a picture
and wrote two number models to help keep track of his thinking.

> Mr. Riley has 2 packs of pencils with 5 pencils in each pack.
> He gives 4 of the pencils to his students.
> How many pencils does he still have?

$$2 \times 5 = 10$$
$$10 - 4 = 6$$

Do Jeremiah's number models fit the number story? Explain your answer.

_____

_____

_____

_____

_____

⑧ There are 5 giant balloons in a pack.

**a.** How many balloons are in 5 packs?
You may draw a picture to help you solve.

Circle the number model that fits the story.
$5 + 5 = ?$      $5 \times 5 = ?$

Answer: _____
(unit)

**b.** Explain how you solved Problem 8a. _____

_____

_____

# Unit 2 Assessment (continued)

⑨ You have 2 rows of chairs with 9 chairs in each row.
How many chairs do you have in all?

   **a.** Draw an array on the dot grid to match the story.

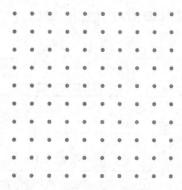

   **b.** Circle the number model that fits the story.

     $2 \times 9 = ?$      $2 + 9 = ?$

     There are _____ in all.
                                      (unit)

⑩ Share 20 marbles equally among 5 friends.
Draw a picture to show how you shared the marbles.

Each friend gets _____.
                                  (unit)
There are _____ left over.
                           (unit)

# Unit 2 Challenge

(1) Lila says that knowing $3 + 7 = 10$ helps her solve this problem on her calculator:

   Enter 423. Change it to 480. How? _____

Explain how Lila might use the basic fact.

_____

_____

_____

(2) Read the number story and circle the pair of number models that fit the story. Then solve.

   Mrs. Ball equally shared 30 markers among 3 groups.
   Mike's group found 6 more markers. How many markers
   does Mike's group have now? You may draw a picture to help.

Circle the pair of number models that best fit the story.

| | | | |
|---|---|---|---|
| **A** | $30 \div 3 = 10$ | **B** | $30 \times 3 = 90$ |
| | $10 + 6 = 16$ | | $90 + 6 = 96$ |
| **C** | $30 + 3 = 33$ | **D** | $30 - 3 = 27$ |
| | $33 + 6 = 39$ | | $27 + 6 = 33$ |

Mike's group now has _____ markers.

# Unit 2 Challenge (continued)

(3) You have 18 chairs that you want to arrange in an array.

    **a.** Show 3 different ways you could do this on the dot grid at the right. Write number models for each array.

        Number models: _____

                         _____

                         _____

    **b.** Can you make an 18-chair array with 5 rows? Explain.

_____

_____

_____

_____

(4) Harrison is making a Frames-and-Arrows problem.
His first two frames show 3 and then 6.
Write a rule that Harrison might be using. Then fill in the frames.

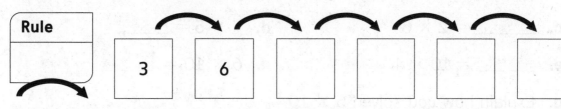

Write a rule that gives different numbers for the other frames.
Then fill in the frames.

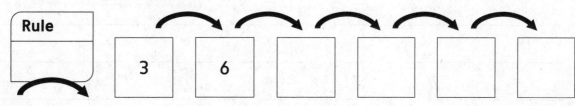

# Unit 2 Cumulative Assessment

(1) Record the time shown on each clock.

**a.**

**b.**

**c.**

_____:_____      _____:_____     _____:_____

(2) Draw the hands to show the times.

**a.**   8:15

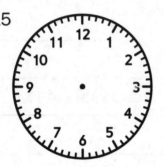

**b.**   8:30

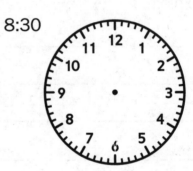

(3) Solve.

**a.** $2 \times 7 =$ _____      **b.** _____ $= 5 \times 5$

**c.** _____ $= 2 \times 6$      **d.** $7 \times 5 =$ _____

**e.** _____ $= 10 \times 4$      **f.** $6 \times 10 =$ _____

**g.** Explain how you solved $6 \times 10$.

_____

_____

_____

_____

# Unit 2 Cumulative Assessment (continued)

④ Round each number to the nearest 10.
You may use open number lines to help.

**a.** 42 _____

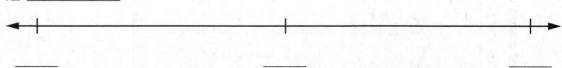

_____           _____           _____

**b.** 88 _____

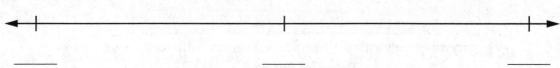

_____           _____           _____

**c.** Explain how you rounded 88 to the nearest 10.

_____

_____

_____

⑤ Round each number to the nearest 100.
You may use open number lines to help.

**a.** 490 _____

_____           _____           _____

**b.** 520 _____

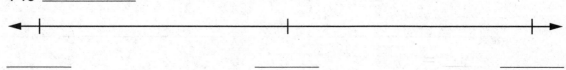

_____           _____           _____

# Unit 2 Cumulative Assessment (continued)

⑥ Use the information in the bar graph to answer the questions below.

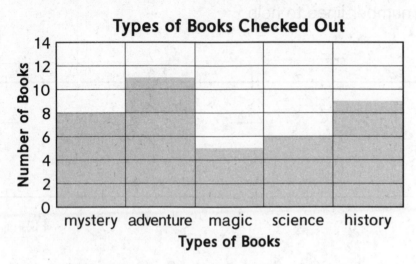

**Types of Books Checked Out**

**a.** How many mystery and adventure books were checked out all together? _____

**b.** How many more adventure books were checked out than science books? _____

**c.** How many books were checked out in all? _____

**d.** Explain how you solved for the number of books checked out in all.

_____

_____

_____

# Unit 3 Self Assessment

Put a check in the box that tells how you do each skill.

| Skills | I can do this on my own and explain how to do this. | I can do this on my own. | I can do this if I get help or look at an example. |
|---|---|---|---|
| ① Complete "What's My Rule?" tables and find rules.   MJ1 65 | | | |
| ② Add 3-digit numbers.   MJ1 69 71 | | | |
| ③ Subtract 3-digit numbers.   MJ1 73 75 | | | |
| ④ Use +, −, ×, and ÷ to write equivalent names for numbers.   MJ1 96 | | | |
| ⑤ Use the adding-a-group strategy to solve facts.   MJ1 90–91 | | | |
| ⑥ Use the subtracting-a-group strategy to solve facts.   MJ1 93–94 | | | |

# Unit 3 Assessment

Complete the tables. Write your own number pair in the last row of each table.

① in ↓

| Rule |
|------|
| Add 4 |

out

| in | out |
|----|-----|
| 14 |     |
|    | 12  |
| 7  |     |
| 15 |     |
|    | 32  |

② in ↓

| Rule |
|------|
|      |

out

| in | out |
|----|-----|
| 10 | 1   |
| 13 | 4   |
| 30 |     |
|    | 43  |
|    | 13  |

For each problem, use rounding to estimate and then solve.
Use your estimate to check whether your answer makes sense.
Show your work.

③  **a.** Estimate: _____

**b.**
```
  1 6 9
+   2 8
```

④  **a.** Estimate: _____

**b.**
```
  8 2
- 3 6
```

**c.** Does your answer make sense? Explain.

_____

_____

# Unit 3 Assessment (continued)

⑤  **a.** Estimate: _____

    **b.**    3 8 6
       + 1 4 5

    **c.** Does your answer make sense? Explain.

_____

_____

⑥  **a.** Estimate: _____

    **b.**    2 9 3
       −  8 5

⑦  Use the tally chart and the key to complete the picture graph.

### 3rd Grade Milk Choices

| Kind of Milk | Number of Children |
|---|---|
| white | ̶H̶H̶ ̶H̶H̶ ̶H̶H̶ ̶H̶H̶ ̶H̶H̶ |
| chocolate | ̶H̶H̶ ̶H̶H̶ ̶H̶H̶ ̶H̶H̶ ̶H̶H̶ ̶H̶H̶ ̶H̶H̶ ̶H̶H̶ |
| no milk | ̶H̶H̶ ̶H̶H̶ |

### 3rd Grade Milk Choices

white    ☐ ☐ ☐ ☐ ☐

chocolate

no milk

Key: ☐ = 5 children

# Unit 3 Assessment (continued)

**8** Use the turn-around rule to solve and draw arrays for each fact.

**a.** $5 \times 7 =$ _____          $7 \times 5 =$ _____

**b.** _____ $= 10 \times 3$          _____ $= 3 \times 10$

**c.** How does drawing arrays for these fact pairs help you understand the turn-around rule?

_____

_____

_____

_____

# Unit 3 Assessment (continued)

⑨ Write a number sentence to match each array.

**a.** • • •
    • • •

Number sentence: _____

**b.** • •
    • •

Number sentence: _____

**c.** Which array, a or b, in Problem 9 shows a multiplication square? Explain.

_____

_____

_____

⑩ Li does not know the answer to 6 × 4.
She does know that 5 × 4 = 20, so she uses it as a helper fact.
Li starts by drawing this array for 5 × 4 = 20:

× × × ×
× × × ×
× × × ×
× × × ×
× × × ×

Show on the picture and explain how Li can use this array to help her figure out 6 × 4.

_____

_____

# Unit 3 Challenge

① Elias likes to skip count equal groups when he is multiplying. He has to solve 10 × 4.

   **a.**  10 × 4 means _____ groups of _____

       4 × 10 means _____ groups of _____

   **b.**  How are 10 × 4 and 4 × 10 alike?

      _____

      _____

   **c.**  Would it be easier for Elias to skip count 4 groups of 10 or 10 groups of 4? Explain.

      _____

      _____

      _____

# Unit 3 Challenge (continued)

② Logan wants to solve 8 × 7. She knows 10 × 7 = 70.

   **a.** 10 × 7 means _____ groups of _____

       8 × 7 means _____ groups of _____

   **b.** Logan uses the subtracting-a-group strategy with 10 × 7 to help her figure out 8 × 7. Use numbers, pictures, or words to explain what Logan did.

        8 × 7 = _____

# Unit 3 Open Response Assessment
## Finding a Mistake in a Subtraction Problem

Mia wants to solve this problem: 552 − 153 = ?
She begins by making an estimate.

Estimate: $\underline{550 - 150 = 400}$

Then she uses expand-and-trade subtraction to find an exact answer, but her answer is not close to her estimate. This is her work:

$$
\begin{array}{r}
140 \\
40 \quad 12 \\
552 \rightarrow 500 + \cancel{50} + \cancel{2} \\
- \, 153 \rightarrow \underline{100 + 50 + 3} \\
400 + 90 + 9 = 499
\end{array}
$$

"Oops," said Mia, "I didn't cross out 500 and write 400."
Explain **why** not changing 500 to 400 is a mistake.

(Hint: Use what you know about place value in your answer.)

Copyright © McGraw-Hill Education. Permission is granted to reproduce for classroom use.

# Unit 4 Self Assessment

Put a check in the box that tells how you do each skill.

| Skills | I can do this on my own and explain how to do this. | I can do this on my own. | I can do this if I get help or look at an example. |
|---|---|---|---|
| ① Use a ruler to measure to the nearest $\frac{1}{2}$ inch. **MJ1** 99 | | | |
| ② Organize data on a line plot. **MJ1** 104 | | | |
| ③ Describe a quadrilateral. **MJ1** 112 | | | |
| ④ Find the perimeter of a polygon. **MJ1** 114 | | | |
| ⑤ Find the area of a rectangle. **MJ1** 125 | | | |
| ⑥ Find the area of a rectilinear figure. **MJ1** 131 | | | |

# Unit 4 Assessment

① Measure the line segments to the nearest $\frac{1}{2}$ inch. Write the unit.

_____

about: _____
                                   (unit)

_____

about: _____
                                   (unit)

② Use the data in the tally chart to make a line plot.
Use Xs to show the data on the line plot.

| Lengths of Earthworms to the Nearest $\frac{1}{2}$ Inch | Number of Earthworms |
|---|---|
| $2\frac{1}{2}$ | // |
| 3 | //// |
| $3\frac{1}{2}$ | //// |
| 4 | ~~////~~ |
| $4\frac{1}{2}$ | /// |
| 5 | / |

**Lengths of Earthworms to the Nearest $\frac{1}{2}$ Inch**

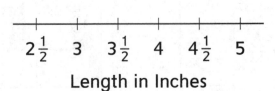

Length in Inches

# Unit 4 Assessment (continued)

③ Xavier is playing *What's My Polygon Rule?*.
He places his polygons this way:

**Fits the Rule**                **Does Not Fit the Rule**

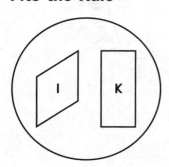 

**a.** Draw a different shape that fits the rule.

**b.** What could Xavier's rule be? Explain how you know.

_____

_____

_____

④ Look at these shapes.

**a.** How are they alike? _____

_____

_____

**b.** How are they different? _____

_____

_____

# Unit 4 Assessment (continued)

⑤ **a.** Trace the boundary of this shape.
Then find the perimeter.
Remember to write the units.

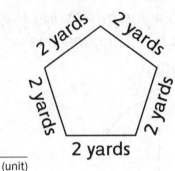

Perimeter: _____

(unit)

**b.** Explain how you figured out the perimeter.

_____

_____

_____

_____

**c.** Which name(s) could be used to name the shape in 5a?
Mark the box next to all that apply.

☐ hexagon       ☐ polygon

☐ pentagon     ☐ quadrilateral

⑥ Find the perimeter and the area of this rectangle.

Key: ☐ = 1 square centimeter

**a.** Perimeter = _____ centimeters

**b.** Area = _____ square centimeters

## Unit 4 Assessment (continued)

⑦ The sewing club made a quilt from 1-foot squares. Molly says the *perimeter* of the quilt is 16 feet and the *area* is 16 square feet.

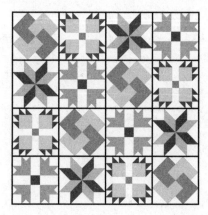

Do you agree with Molly? Explain.

_____

_____

_____

_____

⑧ Find the area of this rectangle.

☐ = 1 square meter

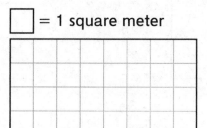

This is a _____-by-_____ rectangle.

Area = _____ square meters

Number sentence: _____ × _____ = _____

# Unit 4 Assessment (continued)

⑨ You draw this card in *The Area and Perimeter Game:*

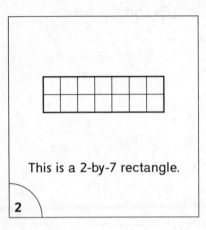

This is a 2-by-7 rectangle.

2

Find the area and the perimeter.

Area: _____ square units

Perimeter: _____ units

⑩ **a.** Partition this rectilinear shape into 2 rectangles.

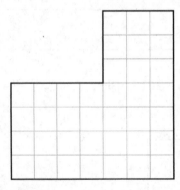

**b.** Find the area of each rectangle.

Area of one rectangle: _____ square units

Area of other rectangle: _____ square units

**c.** Add the areas of your rectangles to find the area of the whole shape.

Area of whole shape: _____ square units

# Unit 4 Challenge

**(1)** Mimi measured this line segment in inches
and says it is about 5 inches long.
Kendall measured the line segment in $\frac{1}{2}$ inches
and says it is about $10\frac{1}{2}$-inches long.

_____

Do you agree with Mimi and Kendall? Explain your answer.

_____

_____

_____

_____

**(2)** What is the smallest number of sides a polygon can have? _____
Draw an example of a polygon with this many sides.

Why are there no polygons with fewer sides?

_____

_____

_____

_____

# Unit 4 Challenge (continued)

③ Your partner draws this card and a "Partner's Choice" card in *The Area and Perimeter Game:*

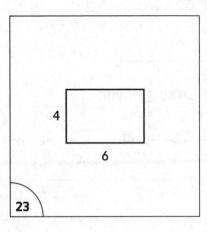

Would you have your partner record the area or the perimeter? Explain.

_____

_____

_____

_____

# Unit 4 Cumulative Assessment

(1) Write a multiplication number sentence for this array.

Number sentence: _____

(2) Draw an array to match this number sentence.

$7 \times 3 = 21$

(3) Fill in the blanks.

    **a.** _____ $\times 2 = 14$

    **b.** $30 = 6 \times$ _____

    **c.** $5 \times$ _____ $= 50$

(4) Monique was solving this problem: $40 \div 5 = ?$
She asked herself, "5 times what number is 40?"
Then she knew the answer. How did Monique figure out the answer?

_____

_____

_____

# Unit 4 Cumulative Assessment (continued)

⑤ Complete.

| in | out |
|----|-----|
| 3 | |
| 5 | |
| | 30 |
| 8 | |
| | |

**Rule**

× 5

⑥ Round each number to the nearest 10 and to the nearest 100.

| | a. Round to the nearest 10. | b. Round to the nearest 100. |
|-----|------------------------------|-------------------------------|
| 247 | | |
| 489 | | |
| 593 | | |
| 303 | | |

**c.** Explain how you rounded 247 to the nearest 100.

_____

_____

_____

# Unit 4 Cumulative Assessment (continued)

⑦ Dakota started her bike ride at 8:25 A.M. She rode until 8:50 A.M.
How many minutes did she ride her bike? Show your work.
You may use your toolkit clock or draw an open number line.

Dakota rode her bike for _____
                                              (unit)

⑧ Use the information in the bar graph to answer the questions below.

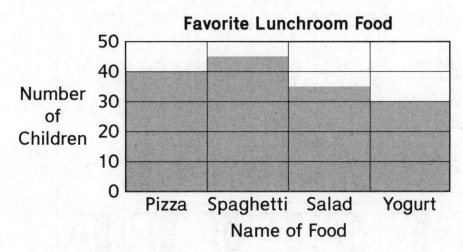

**Favorite Lunchroom Food**

a. How many children voted in all? _____

b. How many more children chose spaghetti than yogurt? _____

c. How many more children voted for salad and yogurt than

   for spaghetti? _____

d. How did you solve 8c? Explain.

_____

_____

_____

_____

# Unit 5 Self Assessment

Put a check in the box that tells how you do each skill.

| Skills | I can do this on my own and explain how to do this. | I can do this on my own. | I can do this if I get help or look at an example. |
|---|---|---|---|
| ① Represent fractions with pictures, words, and numbers. **MJ2** 154 | | | |
| ② Find equivalent fractions using fraction circle pieces. **MJ2** 156–157 | | | |
| ③ Use doubling to help solve other facts. **MJ2** 164–165 | | | |
| ④ Use multiplication squares to help solve other facts. **MJ2** 180–181 | | | |
| ⑤ Identify patterns in products and explain them. **MJ2** 174–175 | | | |
| ⑥ Break apart a factor to make two easier facts. **MJ2** 185–186 | | | |

# Unit 5 Assessment

① Use your fraction circle pieces to complete the table.

| Picture | Words | Number |
|---|---|---|
| Example: The whole is the red piece. | two-eighths | $\dfrac{2}{8}$ |
| The whole is the orange piece. | | |
| The whole is the pink piece. | | |
| The whole is the _____ piece. | two-thirds | |

# Unit 5 Assessment (continued)

② Benjamin turns over these two cards during a game of *Fraction Memory*.

He thinks he found a pair of equivalent fractions.

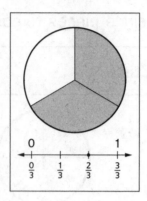 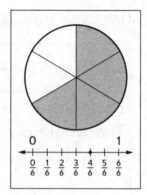

**a.** Do you agree?
Explain your thinking.

_____

_____

**b.** Use your fraction cards to find a different pair of equivalent fractions. Record your two fractions on the lines below.

_____ = _____

③ Complete the table of 4s multiplication facts below.

| Fact | Product |
|------|---------|
| 1 × 4 | |
| 2 × 4 | |
| 3 × 4 | |
| 4 × 4 | |

What patterns do you notice in the products?

_____

_____

_____

## Unit 5 Assessment (continued)

④ For each fact below:
- Record a helper fact.
- Use your helper fact and either add or subtract a group.
- Use words, numbers, or pictures to show your thinking.
- Write the product.

**a.** $3 \times 6 = ?$

Helper fact: _____ × _____ = _____

How I can use the helper fact:


$3 \times 6 =$ _____

**b.** $9 \times 7 = ?$

Helper fact: _____ × _____ = _____

How I can use the helper fact:


$9 \times 7 =$ _____

⑤ Jan is playing a round of *Salute!*. The dealer says 24.
Her partner has an 8 on his forehead.

**a.** What number does Jan have? _____

**b.** Write a multiplication number sentence
and a division number sentence for this problem.

_____     _____

**c.** How do your number sentences show the same *Salute!* round?

_____

_____

# Unit 5 Assessment (continued)

**⑥** Divide the circle below into 4 equal-size parts. Shade and label one part with a fraction.

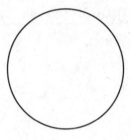

**⑦** Manuel is trying to solve $6 \times 8$.

He sketches a rectangle to help him think about how to break apart the numbers so that the fact is easier to solve. Here is his sketch:

Use numbers or words to explain how Manuel can use his sketch to solve $6 \times 8$.

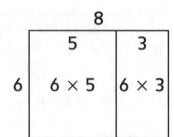

_____

_____

_____

$6 \times 8 =$ _____

**⑧** Tom and Liz are working together to solve $7 \times 8$.

• Tom says: "I think $7 \times 7$ will work as our helper fact."

• Liz says: "I think $8 \times 8$ will work as our helper fact."

With whom do you agree? Explain.

_____

_____

# Unit 5 Challenge

① Explain two different ways you could use doubling to solve $6 \times 8 = ?$. You may draw rectangles to help.

**a.** One way:

Helper fact: _____ × _____ = _____

How I did it:

_____

_____

**b.** Another way:

Helper fact: _____ × _____ = _____

How I did it:

_____

_____

# Unit 5 Challenge (continued)

② Niko is trying to solve $6 \times 9 = ?$.

He sketches a rectangle with side lengths of 6 and 9 to help him think about how he could break it apart to make it easier to solve.

**a.** Show one way Niko could break his rectangle apart.

Record number models to show how he can use the pieces to solve $6 \times 9$.

_____

_____

**b.** Show another way Niko could break his rectangle apart.

Record number models to show how he can use the pieces to solve $6 \times 9$.

_____

_____

**c.** Suppose Niko wants to break his rectangle into 3 parts. Show one way he could do this.

Record number models to show how he can use the pieces to solve $6 \times 9$.

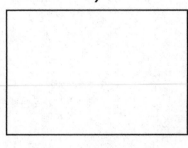

_____

_____

# Unit 5 Open Response Assessment
## Using Multiplication Facts Strategies

Sonja is learning how to use more efficient strategies for multiplication.
She learned about adding or subtracting a group, doubling, and near squares.

She used the adding-a-group strategy to solve $6 \times 7 = ?$. She explained:

> "I will use the helper fact $5 \times 7$. I know that $5 \times 7 = 35$.
> I can add one more group of 7 to 35 to get 42.
> I now have 6 groups of 7, so I know $6 \times 7 = 42$."

(1) Use a picture to show how Sonja solved the problem.
Explain how your picture matches Sonja's explanation.

# Unit 5 Open Response Assessment (continued)

② Choose at least one other efficient multiplication strategy, such as doubling or near squares, to solve $6 \times 7 = ?$. Use pictures and words to show how you solved the problem.

(*Hint*: What helper fact can you use?)

# Unit 6 Self Assessment

Put a check in the box that tells how you do each skill.

| Skills | I can do this on my own and explain how to do this. | I can do this on my own. | I can do this if I get help or look at an example. |
|---|---|---|---|
| ① Subtract multidigit numbers. **MJ2** 189-190 | | | |
| ② Solve number sentences with parentheses. **MJ2** 207 | | | |
| ③ Use order of operations rules to solve number sentences. **MJ2** 212 | | | |
| ④ Write number models to fit number stories. **MJ2** 201 215 | | | |
| ⑤ Identify strategies that can be used to solve facts. **MJ2** 194 | | | |
| ⑥ Solve two-step number stories. **MJ2** 214-215 | | | |

# Unit 6 Assessment

**(1)** Jenny used doubling to solve 6 × 7.
This is what she did:

```
        7
  ┌──────────────┐
  │ 3   3 × 7 = 21│
6 ├──────────────┤
  │ 3   3 × 7 = 21│
  └──────────────┘
```

$6 × 7 = 3 × 7 + 3 × 7$
$6 × 7 = 21 + 21$
$6 × 7 = 42$

**a.** Explain Jenny's work.

_____

_____

_____

**b.** Use doubling to solve 3 × 8.
Draw a picture and write number models.
You may use Jenny's work to help.

# Unit 6 Assessment (continued)

② Fill in the unit box. Then solve.

Unit

a.   7 5 3
    − 3 5 8

b. 331 − 159 = _____

③ In *Baseball Multiplication,* the greater the product from the dice roll, the better the hit.

For each pair of facts below, circle the one that would give a better hit.

a. 6 × 6 or 5 × 8

b. 6 × 9 or 7 × 7

c. 4 × 3 or 2 × 7

④ Show a multiplication strategy that can be used to solve this fact:

8 × 4 = ?

8 × 4 = _____

# Unit 6 Assessment (continued)

⑤ You have 48 stickers and want to divide them equally among 6 small bags. How many stickers do you put into each bag?

- Write a number model to fit the story.
  Use a letter to represent what you want to find out.
  You may complete the diagram below to help.

- Solve the number story.

- Write the number model with your answer to check that your answer makes the number model true.

Letter and what it represents: _____ for _____

| bags | stickers per bag | stickers in all |
|------|------------------|-----------------|
|      |                  |                 |

_____
(number model with letter)

Answer: _____
                              (unit)

_____
(number model with answer)

## Unit 6 Assessment (continued)

⑥ Magi and Katrina solved this number sentence: $3 \times (6 + 2) = ?$
Magi says the answer is 20, and Katrina says the answer is 24.
Who is correct? Explain.

_____

_____

_____

⑦ Andy used the order of operations to solve this number sentence.

$3 + 6 \times 5 = 33$

**Rules for the Order of Operations**

1.  Do operations inside parentheses first. Follow rules 2 and 3 when computing inside parentheses.

2.  Then multiply or divide, in order, from left to right.

3.  Finally add or subtract, in order, from left to right.

Explain Andy's steps for solving the number sentence.

_____

_____

# Unit 6 Assessment (continued)

(8) Solve.

Mrs. Sierra's class has 6 tables with 4 children at each table and a table with 3 children.
How many children are in Mrs. Sierra's class?

Number model: $(6 \times 4) + 3 = C$

**a.** Solve the number story using any strategy. Show your work.

Answer: _____
                                       (unit)

**b.** Explain how the number model fits the story.

_____

_____

_____

_____

# Unit 6 Challenge

① Melanie and Devon subtracted to solve the problem below.

This is Melanie's work.

$$\begin{array}{r} 729 \\ -355 \\ \hline 434 \end{array}$$

| Unit |
|------|
| stars |

This is Devon's work.

$$\begin{array}{r} 6\ \ 12 \\ \cancel{7}\cancel{2}\,9 \\ -355 \\ \hline 374 \end{array}$$

| Unit |
|------|
| rocks |

Who got the correct answer? Who made a mistake?
Explain your thinking.

# Unit 6 Challenge (continued)

② Show how 6 × 8 can be solved using two different efficient multiplication strategies. Show your thinking with number sentences or words.

One way:

Another way:

③ Write a number story to fit this number sentence: $B \times 6 = 42$

$B$ represents _____.

Number story: _____

_____

_____

_____

Solve your number story. Record your answer with units.

_____
(unit)

# Unit 6 Cumulative Assessment

① Solve. You may draw a picture or a diagram.

The animal shelter keeps 5 kittens in each pen.
There are 5 pens. How many kittens are there in all?

Answer: _____
                                    (unit)

Number model: _____

② Fill in the blanks.

| in | out |
|----|-----|
| 8  |     |
|    | 6   |
| 14 |     |
|    | 2   |
|    | 5   |

**Rule**
in →
÷ 2
→ out

| in | out |
|----|-----|
| 4  | 12  |
| 6  | 18  |
| 7  | 21  |
| 9  | 27  |

**Rule**
in →
→ out

# Unit 6 Cumulative Assessment (continued)

③ Fill in the blanks.

**a.** 6 = 6 × _____          **b.** 6 × _____ = 0

**c.** _____ × 0 = 0          **d.** 9 × 5 = 5 × _____

④ Rolando was playing *Salute!* and saw 5 on his partner's forehead. The Dealer said 30.

What is the card on Rolando's forehead? _____

How do you know? _____

_____

_____

⑤ Fill in the blanks.

**a.** 3 × 3 = _____          **b.** _____ = 4 × 4

**c.**    7                      **d.**    9
      × 7                            × 9

**e.** How are the facts in Problems 5a–5d alike?

_____

_____

_____

## Unit 6 Cumulative Assessment (continued)

⑥ Fill in the blanks.

**a.** 30 ÷ 3 = _____          **b.** 4 = _____ ÷ 4

**c.** 5 = 35 ÷ _____          **d.** 50 ÷ 10 = _____

⑦ Draw a picture and use words to explain why 2 × 8 = 8 × 2.

⑧ **a.** Write a name for each quadrilateral.

_____   _____   _____

**b.** Circle one of the shapes above.

Write a different name for that shape. _____

# Unit 6 Cumulative Assessment (continued)

⑨ **a.** Shade the circle on this fraction card to show $\frac{1}{2}$.

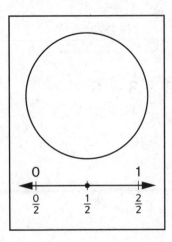

**b.** On this fraction card, partition and shade the circle to show a fraction that is equivalent to $\frac{1}{2}$ but with a different denominator. You may use your fraction cards to help.

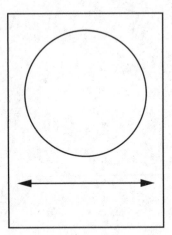

Name the fraction: _____

**c.** How do you know the fractions are equivalent?

_____

_____

# Unit 6 Cumulative Assessment (continued)

⑩ **a.** The mass of a basketball is about 625 grams.
The mass of a baseball is about 142 grams.

About how many more grams is a basketball than a baseball?

Estimate: _____

Answer: about _____
                                            (unit)

**b.** Explain how you know your answer makes sense.

_____

_____

_____

_____

⑪ You draw this card in *The Area and Perimeter Game:*

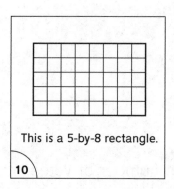

This is a 5-by-8 rectangle.

10

Find the area and the perimeter.

Area: _____ square units           Perimeter: _____ units

# Unit 6 Cumulative Assessment (continued)

⑫  **a.** Fill in the bar graph for the top four sports picked by third graders at Harvey School.

| Sport | Number of Votes |
|---------|-----------------|
| soccer | 30 |
| baseball | 30 |
| dance | 25 |
| swimming | 15 |

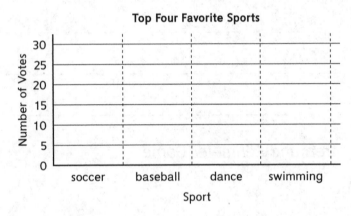

**Top Four Favorite Sports**

**b.** How many more votes did baseball get than dance?

_____

**c.** How many votes are there in all? _____

**d.** Compare the number of votes for soccer to the number of votes for dance and swimming together.
How many more votes were for dance and swimming together than for soccer?

_____

# Unit 6 Cumulative Assessment (continued)

⑬ Measure the length of the line segment to the nearest half inch and nearest centimeter.

_____

about _____ inches          about _____ centimeters

⑭ Measure the side lengths of this rectangle to the nearest inch and label them.

This is a _____ by _____ rectangle.
                    (unit)                              (unit)

Area: _____ square inches

Number model: _____

# Unit 6 Cumulative Assessment (continued)

⑮ Circle all the names for this shape.

rectangle              square              quadrilateral

parallelogram          triangle            rhombus

⑯ Solve each fact.
Write another fact next to each using the turn-around rule.

**a.** 2 × 3 = _____        _____

**b.** _____ = 5 × 7        _____

**c.**    8
        × 2                    _____

**d.**      6
        × 1 0                  _____

# Unit 7 Self Assessment

Put a check in the box that tells how you do each skill.

| Skills | I can do this on my own and explain how to do this. | I can do this on my own. | I can do this if I get help or look at an example. |
|---|---|---|---|
| ① Measure and estimate liquid volumes. **MJ2** 218 | | | |
| ② Solve number stories about measuring mass and liquid volume. **MJ2** 224-225 | | | |
| ③ Compare fractions using fraction strips or number lines. **MJ2** 227, 235 | | | |
| ④ Find equivalent fractions using fraction circles, fraction strips, or number lines. **MJ2** 243 | | | |
| ⑤ Record fraction comparisons with >, <, and =. **MJ2** 227 | | | |
| ⑥ Represent and locate fractions on number lines. **MJ2** 241 | | | |

# Unit 7 Assessment

① Circle the container that is most likely to hold about 1 liter of liquid.

    cup              water bottle            bucket

Solve each measurement number story in Problems 2–4. Show your work.

② Daniel fills these two beakers and pours them into his jar.

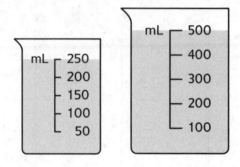

There is no room left in his jar.

What is the liquid volume of his jar?

Answer: about _____ mL (milliliters)

## Unit 7 Assessment (continued)

③ Allison fills a beaker with 1,000 milliliters of water.
Then she pours some of the water from the beaker to fill a glass.
There are 300 milliliters of liquid left in the beaker.

What is the liquid volume of the glass?

Answer: about _____ mL (milliliters)

④ One crayon has a mass of about 5 grams.
What is the mass of 12 crayons together?

Answer: about _____ grams

# Unit 7 Assessment (continued)

**(5)** Kristen uses her fraction strips to compare $\frac{1}{3}$ and $\frac{1}{4}$.

| $\frac{1}{3}$ | $\frac{1}{3}$ | $\frac{1}{3}$ |
|---|---|---|

| $\frac{1}{4}$ | $\frac{1}{4}$ | $\frac{1}{4}$ | $\frac{1}{4}$ |
|---|---|---|---|

Kristen writes this number sentence: $\frac{1}{3} < \frac{1}{4}$

Do you agree with Kristen? _____

Use Kristen's fraction strips to help explain your answer.

_____

_____

_____

**(6)** Partition the number line into fourths and label each tick mark.

You may use the fraction strip to help.

| $\frac{1}{4}$ | $\frac{1}{4}$ | $\frac{1}{4}$ | $\frac{1}{4}$ |
|---|---|---|---|

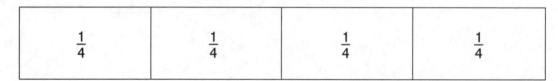

$\frac{0}{4}$

$\frac{\square}{4}$

## Unit 7 Assessment (continued)

⑦ How far did the triangle move? Record the fraction.

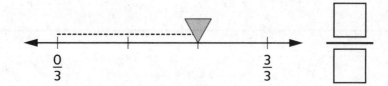

$\frac{0}{3}$          $\frac{3}{3}$

⑧ Write >, <, or = to make the number sentences true.
The whole is the same for each fraction.
You may use your fraction tools.

> < means *is less than*
> \> means *is greater than*
> = means *is equal to*

a. $\frac{1}{8}$ _____ $\frac{1}{2}$

b. $\frac{3}{4}$ _____ $\frac{3}{6}$

c. $\frac{4}{2}$ _____ $\frac{3}{2}$

d. $\frac{4}{8}$ _____ $\frac{2}{4}$

e. Show how you can compare $\frac{4}{8}$ and $\frac{2}{4}$ using the number lines below.

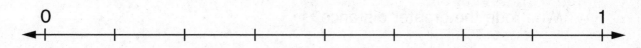

0                                                    1

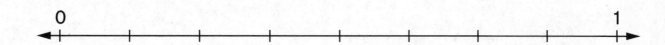

0                                                    1

# Unit 7 Assessment (continued)

**⑨ a.** Fill in the missing thirds on the number line.

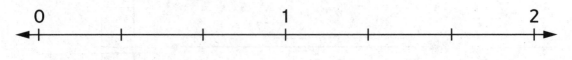

**b.** Draw a point at $\frac{4}{3}$.

**c.** Is $\frac{4}{3}$ greater than, less than, or equal to 1? _____

How do you know? _____

_____

**⑩** Solve the fraction stories. Show your work.
Use fraction circles, fraction strips, number lines, or drawings.

**a.** Ron rode his bike $\frac{1}{6}$ of a mile.

Tammy rode her bike $\frac{1}{8}$ of a mile.

Who rode the greater distance?

Answer: _____

**b.** Four friends share 3 oranges equally.

What fraction of an orange does each friend get?

Answer: _____
(unit)

## Unit 7 Assessment (continued)

**(11)** **a.** What fraction is this fraction strip showing?

| $\frac{1}{4}$ | $\frac{1}{4}$ | $\frac{1}{4}$ |

_____ of a fraction strip

**b.** Partition this fraction strip to show halves.
Label with fractions.

**(12)** Draw a line from each number sentence to the picture that matches it.

$\frac{1}{2} > \frac{1}{4}$

| $\frac{1}{6}$ | $\frac{1}{6}$ | $\frac{1}{6}$ | $\frac{1}{6}$ | $\frac{1}{6}$ | $\frac{1}{6}$ |

| $\frac{1}{3}$ | $\frac{1}{3}$ | $\frac{1}{3}$ |

$\frac{2}{3} < \frac{3}{3}$

$\frac{0}{2}$      $\frac{2}{2}$      $\frac{4}{2}$

$\frac{0}{4}$      $\frac{4}{4}$      $\frac{8}{4}$

$\frac{2}{6} = \frac{1}{3}$

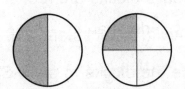

$\frac{3}{2} > \frac{3}{4}$

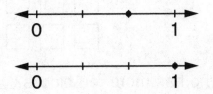

0      1

0      1

# Unit 7 Assessment (continued)

**13** Alexander made a mistake when he labeled $\frac{1}{2}$ on the number line below.

He crossed out his mistake but needs help to fix it.

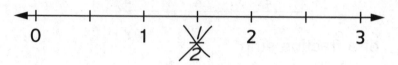

**a.** Explain Alexander's mistake.

_____

_____

_____

**b.** Label $\frac{1}{2}$ on the number line.

**14** **a.** Four people share 8 pennies. Circle each person's share.

◯ ◯ ◯ ◯ ◯ ◯ ◯ ◯

How many pennies does each person get? _____ pennies

Write the fraction of the total number of pennies that each

person gets. _____ of the pennies

**b.** Sai and Anika each have 6 blocks.

$\frac{2}{6}$ of Sai's blocks are red.

$\frac{4}{6}$ of Anika's blocks are red.

Shade the blocks to show Sai's and Anika's red blocks.

Sai's blocks        Anika's blocks

☐ ☐ ☐        ☐ ☐ ☐
☐ ☐ ☐        ☐ ☐ ☐

Who has more red blocks? _____

# Unit 7 Challenge

① **a.** Mark and label the points $\frac{3}{4}$, $\frac{7}{4}$, and $\frac{10}{4}$ on the number line.

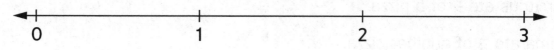

```
0        1        2        3
```

**b.** Write <, >, or = to make the number sentences true.

Use the number line above to help.

$\frac{7}{4}$ _____ 2          $\frac{10}{4}$ _____ 2

② Felipe shared his collection of 12 baseball cards equally with his brother. Write at least 3 different equivalent fractions that name each boy's share of the cards.

_____  _____  _____

③ Write <, >, or = to make the number sentences true.

You may use fraction tools to help.

**a.** $\frac{3}{4}$ _____ $\frac{4}{8}$          **b.** $\frac{6}{4}$ _____ $\frac{3}{2}$

**c.** $\frac{2}{6}$ _____ $\frac{7}{8}$          **d.** $\frac{3}{4}$ _____ $\frac{6}{8}$

**e.** Choose a fraction tool to help you compare $\frac{3}{4}$ and $\frac{4}{8}$. Draw a picture to show what you did.

# Unit 7 Open Response Assessment

## Halves of a Whole

Demitrius ate $\frac{1}{2}$ of a pizza.

Emma ate $\frac{1}{2}$ of another pizza.

Demitrius said that he ate more pizza than Emma,
but Emma said they both ate the same amount.

Use words and pictures to show that Demitrius could be right.

Use words and pictures to show that Emma could be right.

# Unit 8 Self Assessment

Put a check in the box that tells how you do each skill.

| Skills | I can do this on my own and explain how to do this. | I can do this on my own. | I can do this if I get help or look at an example. |
|---|---|---|---|
| ① Measure lengths of objects to the nearest $\frac{1}{4}$ inch. **MJ2** 252 | | | |
| ② Solve multiplication and division extended facts. **MJ2** 255-256 | | | |
| ③ Know most multiplication facts. **MJ2** 298-301 | | | |
| ④ Solve equal sharing problems. **MJ2** 266 | | | |
| ⑤ Describe attributes of the faces and bases of prisms. **MJ2** 272 | | | |
| ⑥ Name factor pairs for products. **MJ2** 259 | | | |

# Unit 8 Assessment

①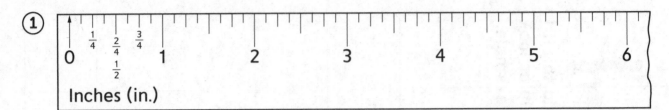

**a.** Make a dot at $4\frac{1}{2}$ inches from 0. Label it with the letter *A*.

**b.** Make a dot at $3\frac{3}{4}$ inches from 0. Label it with the letter *B*.

**c.** Make a dot at $5\frac{1}{4}$ inches from 0. Label it with the letter *C*.

② Measure the line segment below to the nearest $\frac{1}{4}$ inch.

_____

about _____ in.

③ Write a helper fact and use it to help you solve.

**a.** $2 \times 90 =$ _____

Fact I used to help:

_____

**b.** $60 \times 5 =$ _____

Fact I used to help:

_____

**c.** $6 \times 80 =$ _____

Fact I used to help:

_____

Use the helper fact to help you fill in the missing factors.

**d.** Helper fact: $6 \times 2 =$ _____

$60 \times$ _____ $= 120$

**e.** Helper fact: _____ $= 4 \times 5$

$200 =$ _____ $\times 5$

**f.** Helper fact: $7 \times 5 =$ _____

_____ $\times 50 = 350$

## Unit 8 Assessment (continued)

④ Write in factor pairs to make the number sentences true.

_____ × _____ = 16

24 = _____ × _____

_____ × _____ = 40

⑤ Three friends want to share $42. They have $10 bills and $1 bills. They can exchange larger bills for smaller bills if they need to. Write a number model. Use numbers or pictures to show how you solved the problem.

The letter _____ stands for _____.

_____

(number model with letter for unknown)

Answer: Each friend gets $_____.

# Unit 8 Assessment (continued)

⑥ Here is a *Factor Bingo* game mat. You draw a 5 card.

Circle at least two products with a factor of 5.

| 8 | 24 | 51 | 64 | 80 |
|----|----|----|----|----|
| 76 | 32 | 10 | 48 | 17 |
| 55 | 16 | 42 | 54 | 90 |
| 28 | 40 | 36 | 15 | 66 |
| 63 | 72 | 9 | 56 | 12 |

⑦ Explain why the shape in this picture is a rectangular prism.

_____

_____

_____

# Unit 8 Challenge

① Suppose 4 friends want to share $86. They have $10 bills and $1 bills.
Show or explain how much money each friend would get.
Be sure to describe each step of how you shared the $86.

Number model: _____

Each friend gets _____.

② Here is a game mat for *Speed Factor Bingo.*

| 8 | 24 | 51 | 64 | 80 |
|---|----|----|----|----|
| 76 | 32 | 10 | 48 | 17 |
| 55 | 16 | 42 | 54 | 90 |
| 28 | 40 | 36 | 15 | 66 |
| 63 | 72 | 9 | 56 | 12 |

In *Speed Factor Bingo,* a player draws a number card and covers all the products that have that number as a factor.

Name a factor card that would allow a player to get a bingo in one turn.

_____

Draw a line through the row, column, or diagonal to show the bingo.

# Unit 8 Challenge (continued)

③ Julian traced the bases and other faces of a pattern-block prism.

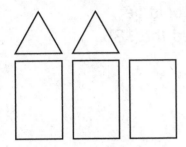

Circle the picture of the prism that matches his tracings.

Name the shapes of its bases. _____

Name the shapes of its other faces. _____

Louis says this is a picture of a rectangular prism.
Explain why you agree or disagree.

_____

_____

# Unit 8 Cumulative Assessment

For each story:

- Write a number model. Use a letter for what you want to find out. You may complete the diagram to help.

- Solve. Then write the number model with your answer to check your work.

① Laurie bought 7 boxes of pencils. There were 8 pencils per box.

How many pencils did she buy in all?

| Boxes | Pencils in each box | Pencils in all |
|-------|---------------------|----------------|
|       |                     |                |

The letter _____ represents _____.

_____
(number model with letter)

Laurie bought _____.
                                         (unit)

_____
(number model with answer)

# Unit 8 Cumulative Assessment (continued)

② The art teacher shared 40 balls of yarn equally among the 10 children in the art club. How many balls of yarn did each child get?

| Children | Balls of yarn per child | Balls of yarn in all |
|---|---|---|
|  |  |  |

The letter _____ represents _____.

_____
(number model with letter)

Each child got _____.
                                    (unit)

_____
(number model with answer)

③ Fill in the blanks.

a. $7 \times$ _____ $= 28$          b. _____ $= 8 \times 7$

c. _____ $\times 9 = 27$          d. $42 =$ _____ $\times 7$

e. If $9 \times$ _____ $= 36$, then $36 \div 9 =$ _____.

f. If _____ $\times 7 = 63$, then $63 \div 7 =$ _____.

g. If $8 \times$ _____ $= 72$, then $72 \div$ _____ $= 8$.

# Unit 8 Cumulative Assessment (continued)

④ A.J. used the break-apart strategy to solve $7 \times 9$ by breaking 9 into the easier numbers 5 and 4. See her picture below.

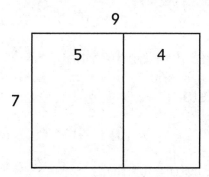

Use A.J.'s easier numbers and drawing to write number models that she can use to help solve $7 \times 9$.

$7 \times 9 =$ _____

⑤ Fill in the blanks.

**a.** $9 \div$ _____ $= 3$

**b.** $25 \div 5 =$ _____

**c.** _____ $\div 4 = 4$

**d.** $49 \div 7 =$ _____

# Unit 8 Cumulative Assessment (continued)

⑥ Nathan has 5 bags of marbles.
Each bag has 4 yellow marbles and 6 red marbles.

How many marbles does Nathan have in all?

The letter *M* represents the number of marbles that Nathan has.

**a.** Underline the number model that fits the story.

$5 \times 4 + 6 = M$        $5 \times (4 + 6) = M$        $(5 + 4) \times 6 = M$

**b.** Solve the number story. You may draw a picture to help.

Answer: _____
                       (unit)

**c.** Write the number model with your answer to check your work.

_____

# Unit 8 Cumulative Assessment (continued)

⑦ Cross out the names that do not belong.

Add at least two more names with parentheses that belong in the name-collection box.

> **18**
>
> $(3 \times 4) + 6$      $3 \times (4 + 6)$
>
> $(64 \div 8) + 10$    $(18 + 8) \times 0$
>
> $3 \times (36 \div 6)$

⑧ For each problem, make an estimate and solve. Check to make sure your answer makes sense.

Unit

**a.** Estimate: _____

$$\begin{array}{r} 5\ 3\ 9 \\ +\ 3\ 5\ 8 \\ \hline \end{array}$$

**b.** Estimate: _____

$847 - 648 =$ _____

# Unit 8 Cumulative Assessment (continued)

⑨ Partition the circle into 8 equal parts. Label each part.

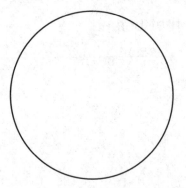

Shade $\frac{3}{4}$ of the circle.

Write two fractions that name the **unshaded** part of the circle.

_____    _____

⑩ Write the time shown on the clocks below.

**a.**

**b.**

_____                    _____

**c.** Draw the hour and minute hands to show the time 15 minutes before 9:27.

What time does the clock show? _____

# Unit 8 Cumulative Assessment (continued)

⑪ Frank practiced his drums for 50 minutes.

He started playing at 4:18 P.M. What time did he finish?

He finished at _____ P.M.

⑫ Ahmed has 900 milliliters (mL) of water in his watering can.
One jar holds 379 mL of water and the other holds 483 mL of water.
How much water does Ahmed need to fill both jars?

**a.** Estimate: _____

Answer: _____
(unit)

**b.** Does Ahmed have enough water to fill both jars? _____

Did you need to find an exact answer to decide whether Ahmed has enough water? Explain.

_____

_____

_____

_____

# Unit 8 Cumulative Assessment (continued)

⑬ You draw this card in *The Area and Perimeter Game*:

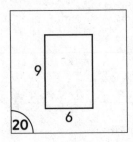

a. Find the area and the perimeter.

Area: _____ square units

Perimeter: _____ units

b. Explain how you found the area.

_____

_____

⑭ Addie wants to put a cloth rug in her dollhouse.
The area she wants to cover is 36 square inches.
If Addie wants a square rug, how long and how wide
should she cut the cloth?

Draw a picture of the rug and label the side lengths.

The rug should be cut _____ long and

_____ wide.
     (unit)

What is the perimeter of the rug? _____
                                        (unit)

# Unit 8 Cumulative Assessment (continued)

⑮ The third-grade class is figuring out the area of the floor in the reading space.

Here is a sketch of the reading space:

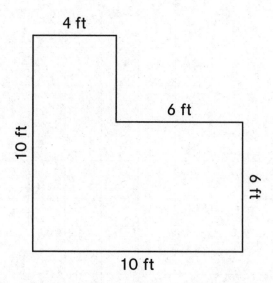

4 ft

6 ft

10 ft

6 ft

10 ft

Draw a line to make two smaller rectangles you can use to find the area.

Show your work. Write the number models you use.

Number models: _____

_____

The area of the reading space is _____.
(unit)

⑯ The perimeter of this rectangle is 24 centimeters.

Label the missing side lengths.

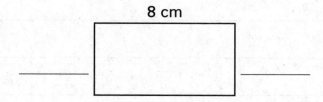

8 cm

# Unit 8 Cumulative Assessment (continued)

⑰ **a.** Draw a rectangle with a perimeter of 16 centimeters.
Then draw a different rectangle with the same perimeter.

Label your rectangles A and B.

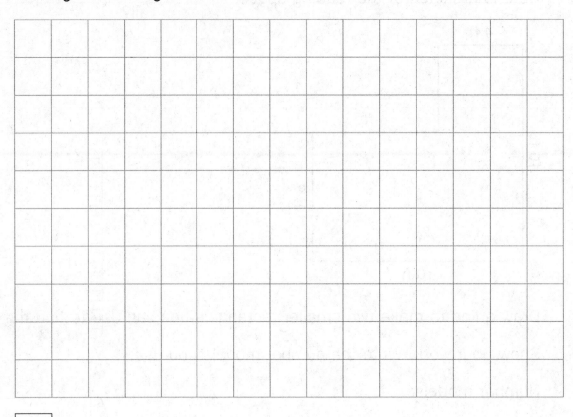

☐ = 1 square cm

**b.** Explain how you know the perimeters for Rectangle A and Rectangle B are 16 centimeters.

_____

_____

_____

**c.** What is the area of Rectangle A? _____
(unit)

**d.** What is the area of Rectangle B? _____
(unit)

# Unit 8 Cumulative Assessment (continued)

⑱ The mass of a baseball is 142 grams.

Greg has one 100-gram mass, one 50-gram mass,
five 10-gram masses, five 5-gram masses, and five 1-gram masses.
What masses could he use to balance the baseball?

_____

_____

⑲ The 1-liter beaker at the right has
350 milliliters of water.
Imani wants to have a full liter of water.
How much more water does she need to add?

She needs _____ more milliliters
of water to make 1 liter.

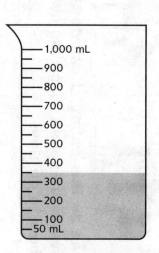

⑳ Tony said $\frac{3}{4}$ of Rectangle A is equal to
$\frac{3}{4}$ of Rectangle B.

Erin said $\frac{3}{4}$ of Rectangle A is not equal to
$\frac{3}{4}$ of Rectangle B.

With whom do you agree? Explain.

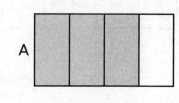

_____

_____

_____

# Unit 9 Self Assessment

Put a check in the box that tells how you do each skill.

| Skills | I can do this on my own and explain how to do this. | I can do this on my own. | I can do this if I get help or look at an example. |
|---|---|---|---|
| ① Know all multiplication facts. **MJ2 298-301** | | | |
| ② Solve all division facts. **MJ2 231** | | | |
| ③ Solve number stories with extended facts. **MJ2 279** | | | |
| ④ Multiply 2-digit numbers by 1-digit numbers. **MJ2 287** | | | |
| ⑤ Tell time to the nearest minute. Check where you wrote the time at the top of all *Math Journal* pages. | | | |
| ⑥ Calculate elapsed time. **MJ2 291** | | | |

# Unit 9 Assessment

① For each number sentence, fill in the blank with a factor from 1 to 10 to make it true.

**a.** $5 \times 7 < 7 \times$ _____

**b.** $4 \times 3 > 4 \times$ _____

**c.** $8 \times 6 <$ _____ $\times$ _____

For Problems 2–4, write a number model with a letter for the unknown. Then solve the problem and write the answer. Write your number model again with the answer to check that your answer makes sense.

② Seven blue jays each have a mass of about 80 grams. What is their total mass?

_____
(number model with letter)

The letter _____ stands for _____.

Seven blue jays have a total mass of about _____ grams.

_____
(number model with answer)

# Unit 9 Assessment (continued)

(3) Together, 80 great blue herons have a mass of about 240 kilograms. One California condor has a mass of about 8 kilograms. About how many 8-kilogram California condors would it take to equal the mass of the group of herons?

_____
(number model with letter)

The letter _____ stands for _____.

It would take about _____ California condors to equal the mass of 80 great blue herons.

_____
(number model with answer)

(4) About how many 50-gram northern cardinals have a mass equal to one 500-gram Atlantic puffin?

_____
(number model with letter)

The letter _____ stands for _____.

It would take about _____ northern cardinals to equal the mass of one Atlantic puffin.

_____
(number model with answer)

# Unit 9 Assessment (continued)

⑤ Use the break-apart strategy to solve each problem.
You may use mental math, drawings, number sentences, or words.
Show your thinking.

**a.**  5 × 32 = _____

**b.**  3 × 61 = _____

# Unit 9 Assessment (continued)

**c.** Ellen drew a rectangle to help solve 8 × 64. Here is her work:

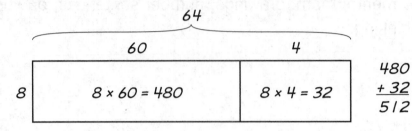

Explain how Ellen solved the problem.

_____

_____

_____

_____

_____

_____

## Unit 9 Assessment (continued)

⑥ It starts raining at 6:40 A.M. and stops at 9:15 A.M.
How long did it rain?
Show your thinking. You may use an open number line,
your toolkit clock, or other representations.

_____ hours _____ minutes

# Unit 9 Challenge

① About how many 10-gram chickadees would equal the mass of six 40-gram cardinals?
Explain your thinking using numbers and words.

About _____ chickadees equal the mass of 6 cardinals.

② Marlena used the break-apart strategy to solve 6 × 76, but she made a mistake:

```
              76
       ┌──────────────┐
          70        6
    ┌─────────┬──────────┐        42
 6  │ 6 × 7 = 42 │ 6 × 6 = 36 │     + 36
    └─────────┴──────────┘        ────
                                   78
```

Explain Marlena's mistake. _____

_____

Use any strategy to correctly solve 6 × 76. Show your work.

6 × 76 = _____

## Unit 9 Challenge (continued)

③ Use the sunrise and sunset information in the chart to figure out the length of day for each city.

| City | Sunrise on 12/16/2016 | Sunset on 12/16/2016 | Length of Day on 12/16/2016 |
|---|---|---|---|
| Stonington, Maine | 7:07 A.M. | 3:59 P.M. | |
| Brownsville, Texas | 7:13 A.M. | 5:43 P.M. | |

What is the difference between the lengths of day for the two cities?

_____

# Unit 9 Open Response Assessment
## Factor Patterns

1. Explore what happens to the product when you double a factor. For example, multiply 4 × 5. Then double the 4 and multiply 8 × 5. Then begin with 4 × 5 again, double the 5, and multiply 4 × 10. Show your work.

2. Try doubling one of the factors in other multiplication facts. Show your work. Describe a pattern that you see when you double a factor.

3. Based on your work with doubling factors, predict what will happen when you triple a factor. Explain how you would convince someone that your prediction will always work for any multiplication fact.

# Beginning-of-Year Assessment

The Beginning-of-Year Assessment can be used to gauge children's readiness for the content they will encounter early in third grade. This allows you to plan your instruction accordingly.

## Goals

The following table provides information about the Common Core State Standards and the Standards for Mathematical Practice assessed in the Beginning-of-Year Assessment.

| CCSS Common Core State Standards | Goals for Mathematical Content (GMC) | Item(s) |
|---|---|---|
| 3.OA.1 | Interpret multiplication in terms of equal groups. | 8b |
| 3.OA.3 | Use multiplication and division to solve number stories. | 8a |
| | Model number stories involving multiplication and division. | 8b |
| 3.OA.8 | Solve 2-step number stories involving two of the four operations. | 7 |
| 3.NBT.2 | Add within 1,000 fluently. | 3a, 4, 5d, 6, 7, 9a, 9b |
| | Subtract within 1,000 fluently. | 3b, 4, 6, 7, 9a, 9b |
| 3.NF.1 | Understand, identify, and represent unit fractions as 1 part when a whole is divided into *b* equal parts. | 10 |
| 3.MD.1 | Tell and write time. | 1a, 1b, 2a, 2b |
| 3.MD.3 | Organize and represent data on scaled bar graphs and scaled picture graphs. | 5 |
| | Solve 1- and 2-step problems using information in graphs. | 5a–5d |
| 3.G.2 | Partition shapes into parts with equal areas. | 10 |
| | Express the area of each part as a unit fraction of the whole. | 10 |

**Goals for Mathematical Practice (GMP)**

| | | |
|---|---|---|
| SMP1 | Make sense of your problem.   GMP1.1 | 4, 7 |
| SMP2 | Create mathematical representations using numbers, words, pictures, symbols, gestures, tables, graphs, and concrete objects.   GMP2.1 | 2a, 2b, 7, 8b |
| | Make sense of the representations you and others use.   GMP2.2 | 1a, 1b, 2a, 2b, 8a, 8b |
| SMP4 | Use mathematical models to solve problems and answer questions.   GMP4.2 | 4, 5, 7 |
| SMP6 | Explain your mathematical thinking clearly and precisely.   GMP6.1 | 4 |
| SMP7 | Look for mathematical structures such as categories, patterns, and properties.   GMP7.1 | 6, 9b |
| | Use structures to solve problems and answer questions.   GMP7.2 | 6, 9b |
| SMP8 | Create and justify rules, shortcuts, and generalizations.   GMP8.1 | 6 |

# Beginning-of-Year Assessment

① Write the time.

**a.**

**b.**

_____     _____

② Draw minute and hour hands to show each time.

**a.** 1:30

**b.** 4:40

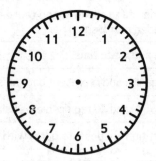

③ Solve.

**a.** 58 + 42 = _____

**b.** 73 − 27 = _____

## Beginning-of-Year Assessment (continued)

④ Together the yellow and blue strings are 57 inches long.
The yellow string is 28 inches long. How long is the blue string?

Write a number model. Use a ? to show the number you need to find.
You may draw a diagram to help.

Number model: _____

Solve the number story. Show your work and explain your thinking.

You may draw an open number line or a picture to help.

The blue string is _____ inches long.

# Beginning-of-Year Assessment (continued)

**Pencils**

⑤ Use the information below to label and complete the graph. Then use the graph to answer the questions.

Tim has 8 pencils.
Cole has 10 pencils.
Vi has 7 pencils.
Dom has 5 pencils.

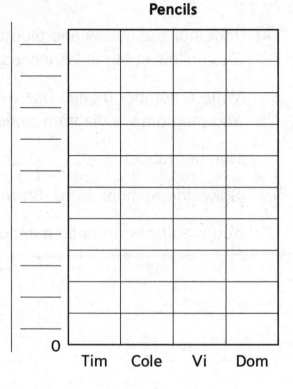

**a.** Who has the most pencils? _____

**b.** Who has the fewest pencils? _____

**c.** How many more pencils does Cole have than Dom? _____ pencils

**d.** How many pencils do the children have all together? _____ pencils

⑥ Find the pattern. Fill in the missing numbers and the rule.

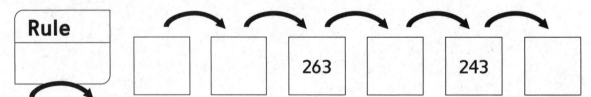

| Rule | | | 263 | | 243 | |

# Beginning-of-Year Assessment (continued)

⑦ Write a number model or number models.
Use a ? to show what you need to find.

To help, you may draw a

| Total | |
| --- | --- |
| Part | Part |

, a

Start ⌒Change End

, or a

| Quantity |
| --- |
| Quantity |
|  Difference |

.

Then use any strategy to solve the number story.

There are 30 children on the playground, and 5 more children join them.
Then 18 children go inside for lunch.
How many children are on the playground now?

Number model(s): _____

There are now _____ children on the playground.

⑧ Derrick drew this array to show 3 groups of 5 pennies.

How many pennies are there?

**a.** There are _____ pennies.

**b.** Write a number sentence to match the array.

_____

# Beginning-of-Year Assessment (continued)

**⑨** **a.** $12 = 6 +$ _____       $4 +$ _____ $= 8$

$16 - 8 =$ _____       _____ $- 7 = 7$

$18 =$ _____ $+ 9$       $5 + 5 =$ _____

**b.** Look at the facts above and think about how they are alike. Then cross out the fact below that does not belong.

$6 = 3 + 3$       $5 + 6 = 11$       $20 - 10 = 10$

**⑩** Partition the circle into two equal parts. Shade one part.

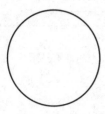

What part of the whole did you shade?

_____

# Mid-Year Assessment

The Mid-Year Assessment covers some of the important concepts and skills presented in *Third Grade Everyday Mathematics*. It should be used to complement the ongoing and periodic assessments that appear within lessons and at the end of each unit.

## Goals

The following table provides information about the Common Core State Standards and the Standards for Mathematical Practice assessed in the Mid-Year Assessment.

| CCSS Common Core State Standards | Goals for Mathematical Content (GMC) | Item(s) |
|---|---|---|
| **3.OA.1** | Interpret multiplication in terms of equal groups. | 1a, 1b |
| **3.OA.2** | Interpret division in terms of equal shares or equal groups. | 2, 3 |
| **3.OA.3** | Use multiplication and division to solve number stories. | 2, 3 |
| | Model number stories involving multiplication and division. | 2, 3 |
| **3.OA.4** | Determine the unknown in multiplication and division equations. | 1b, 4a, 5 |
| **3.OA.5** | Apply properties of operations to multiply or divide. | 1b |
| **3.OA.6** | Understand division as an unknown-factor problem. | 4a, 5 |
| **3.OA.7** | Multiply within 100 fluently. | 1b, 4a, 5, 6a |
| | Know all products of 1-digit numbers × 1, × 2, × 5, and × 10 automatically. | 1b, 4a, 5 |
| | Know all square products of 1-digit numbers automatically. | 6a |
| | Divide within 100 fluently. | 2, 3, 4a, 5 |
| **3.OA.8** | Assess the reasonableness of answers to problems. | 6b, 7b |
| | Solve 2-step number stories involving two of the four operations. | 6a |
| **3.NBT.1** | Use place-value understanding to round whole numbers to the nearest 10. | 9a |
| | Use place-value understanding to round whole numbers to the nearest 100. | 9b |
| **3.NBT.2** | Add within 1,000 fluently. | 7a, 7c, 9a, 9b, 14b |
| | Subtract within 1,000 fluently. | 6a, 8 |
| **3.MD.1** | Tell and write time. | 10a, 10b |
| | Measure time intervals in minutes. | 10c |
| **3.MD.2** | Measure and estimate masses of objects using grams and kilograms. | 11 |
| | Solve 1-step number stories involving mass. | 8 |

# Goals (continued)

| CCSS Common Core State Standards | Goals for Mathematical Content (GMC) | Item(s) |
|---|---|---|
| **3.MD.3** | Organize and represent data on scaled bar graphs and scaled picture graphs. | 14a |
| | Solve 1- and 2-step problems using information in graphs. | 14b |
| **3.MD.5, 3.MD.5a** | Understand that a unit square has 1 square unit of area and can measure area. | 12a |
| **3.MD.5, 3.MD.5b** | Understand that a plane figure completely covered by *n* unit squares has area *n* square units. | 12a |
| **3.MD.6** | Measure areas by counting unit squares. | 12a |
| **3.MD.7, 3.MD.7a** | Find the area of a rectangle by tiling it. | 12a |
| | Show that tiling a rectangle results in the same area as multiplying its side lengths. | 12b |
| **3.MD.7, 3.MD.7b** | Multiply side lengths to find areas of rectangles. | 12b |
| **3.G.1** | Understand that shapes in different categories may share attributes that can define a larger category. | 13a, 13b |
| | Recognize specified subcategories of quadrilaterals. | 13 |
| **3.G.2** | Partition shapes into parts with equal areas. | 12a |

**Goals for Mathematical Practice (GMP)**

| | | |
|---|---|---|
| **SMP1** | Check whether your answer makes sense.  **GMP1.4** | 6b, 7b |
| | Compare the strategies you and others use.  **GMP1.6** | 9c |
| **SMP2** | Create mathematical representations using numbers, words, pictures, symbols, gestures, tables, graphs, and concrete objects.  **GMP2.1** | 1b |
| | Make sense of the representations you and others use.  **GMP2.2** | 1a, 13a |
| **SMP3** | Make sense of others' mathematical thinking.  **GMP3.2** | 7a, 7b |
| **SMP4** | Model real-world situations using graphs, drawings, tables, symbols, numbers, diagrams, and other representations.  **GMP4.1** | 2, 3, 6a, 8, 14a, 14b |
| | Use mathematical models to solve problems and answer questions.  **GMP4.2** | 2, 3, 6a, 14b |
| **SMP5** | Use tools effectively and make sense of your results.  **GMP5.2** | 10c, 11 |
| **SMP6** | Explain your mathematical thinking clearly and precisely.  **GMP6.1** | 4b, 7b, 9c, 13b |
| | Use an appropriate level of precision for your problem.  **GMP6.2** | 7a, 9c |
| | Use clear labels, units, and mathematical language.  **GMP6.3** | 2, 3, 6a, 10c, 13b |
| **SMP7** | Look for mathematical structures such as categories, patterns, and properties.  **GMP7.1** | 4a, ,4b, 5, 13a, 13b |
| | Use structures to solve problems and answer questions.  **GMP7.2** | 1, 4a, 4b, 5, 7b, 12b, 13 |
| **SMP8** | Create and justify rules, shortcuts, and generalizations.  **GMP8.1** | 4a, 4b, 13b |

# Mid-Year Assessment

(1) **a.** Explain how $2 \times 9 = 18$
matches the array.

_____

_____

_____

**b.** Payton wants to use $2 \times 9 = 18$ as a helper fact to solve $3 \times 9 = ?$.
Use the helper fact and the above array to help Payton.
Explain your thinking.

_____

_____

_____

$3 \times 9 =$ _____

(2) Ebony has a 20-inch piece of ribbon.
She cuts it into 4 equal-length pieces.
How long is each piece? Show your work.

Each piece is _____ long.
(unit)

_____
(number model with answer)

# Mid-Year Assessment (continued)

③ Five friends want to share 30 stickers equally.
Draw a picture to show how they could share the stickers.

Each friend gets _____.
                                         (unit)

_____
(number model with answer)

④ Find the rule. Complete the table.

**a.**

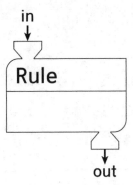

| in | out |
|----|-----|
|    | 20  |
| 7  | 14  |
| 3  | 6   |
| 6  |     |
|    |     |

**b.** How did you figure out the missing rule?

_____

_____

_____

_____

## Mid-Year Assessment (continued)

⑤ Complete the "What's My Rule?" table.

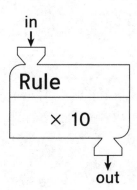

| in | out |
|----|-----|
| 3  |     |
| 10 |     |
|    | 70  |
|    | 20  |

⑥ Davis and his friends have 4 packs of balloons with 4 balloons in each pack. They inflate all of their balloons. Then 3 balloons pop. How many inflated balloons are left?

**a.** Use pictures, numbers, or words to solve the problem. Write number models to show each step.

There are _____ left.
                    (unit)

**b.** How do you know your answer makes sense?

_____

_____

# Mid-Year Assessment (continued)

7) Lily tried to use partial-sums addition to solve this problem.
She thinks she made a mistake.

$$
\begin{array}{r}
643 \\
+291 \\
\hline
800 \\
13 \\
4 \\
\hline
817
\end{array}
$$

**a.** Estimate to check Lily's answer.
Write a number sentence to show how you made your estimate.

_____

**b.** Does Lily's answer make sense? Explain.

_____

_____

_____

**c.** Solve the problem correctly using any strategy. Show your work.

$$
\begin{array}{r}
643 \\
+291 \\
\hline
\end{array}
$$

# Mid-Year Assessment (continued)

⑧ A brown bear has a mass of about 318 kilograms.
A grizzly bear has a mass of about 363 kilograms.
About how much more mass does the grizzly bear have
than the brown bear?
Solve. Show your work.

_____
(number model with ?)

Answer: about _____ kilograms

⑨ **a.** Alice wants to know whether the sum of 686 + 228 is less than
1,000. She rounds both addends to the nearest 10 and uses the
rounded numbers to estimate the sum.

Show how Alice estimated: _____

**b.** Owen rounds both addends to the nearest 100 to estimate the sum.

Show how Owen estimated: _____

**c.** Would you round to the nearest 10 or to the nearest 100
to figure out whether the sum is less than 1,000? Explain.

_____

_____

# **Mid-Year Assessment** (continued)

⑩ Record the time shown on the clocks.

**a.**      A

_____

**b.**      B

_____

**c.** How many minutes pass from the time on Clock A

to the time on Clock B? _____

⑪ Circle at least one item that has a mass of about 1 gram.
Draw a line below at least one item that has a mass of about 1 kilogram.

paper clip       pineapple       centimeter cube

dime       bottle of water       baseball bat

# Mid-Year Assessment (continued)

**12**  **a.** Partition the rectangle into square centimeters.
Then figure out the area.

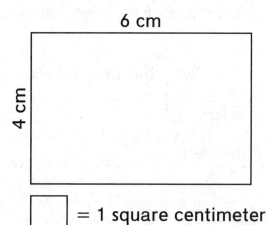

6 cm

4 cm

▢ = 1 square centimeter

Area = _____ square centimeters

**b.** Beatrice found the area by multiplying the side lengths of the
rectangle. Write a number model that she might have used.

_____

**13**  **a.** Circle all the quadrilaterals that are rectangles.

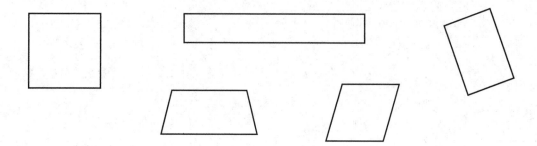

**b.** Explain why the quadrilaterals you circled are rectangles.

_____

_____

_____

# Mid-Year Assessment (continued)

(14) The scouts made blankets for the animal shelter in four different sizes: small, medium, large, and extra large.

**a.** Create a picture graph to show the number of each type of blanket they made.

| Size | Number of Blankets |
|------|-------------------|
| small | 50 |
| medium | 160 |
| large | 110 |
| extra large | 40 |

**Number of Blankets**

small

medium

large

extra large

Key: ☐ = 20 blankets

**b.** The animal shelter only has dogs and cats, so the scouts sent the extra-large blankets to the zoo.
How many blankets did they give to the shelter?
Solve. Show your work.

_____
(number model with ?)

The scouts gave the animal shelter _____ blankets.

# End-of-Year Assessment

The End-of-Year Assessment covers some of the important concepts and skills presented in *Third Grade Everyday Mathematics*. It should be used to complement the ongoing and periodic assessments that appear within lessons and at the end of each unit.

## Goals

The following table provides information about the Common Core State Standards and the Standards for Mathematical Practice assessed in the End-of-Year Assessment.

| CCSS Common Core State Standards | Goals for Mathematical Content (GMC) | Item(s) |
|---|---|---|
| 3.OA.2 | Interpret division in terms of equal shares or equal groups. | 4 |
| 3.OA.3 | Use multiplication and division to solve number stories. | 4, 8, 14 |
| | Model number stories involving multiplication and division. | 4, 8, 14 |
| 3.OA.4 | Determine the unknown in multiplication and division equations. | 3, 4, 5b, 5c, 17a–17d |
| 3.OA.5 | Apply properties of operations to multiply or divide. | 5a–5c, 8 |
| 3.OA.6 | Understand division as an unknown-factor problem. | 3, 4 |
| 3.OA.7 | Multiply within 100 fluently. | 3, 5b, 5c, 7a, 7b, 8, 14, 17a–17e |
| | Know all products of 1-digit numbers $\times 1$, $\times 2$, $\times 5$, and $\times 10$ automatically. | 8, 14 |
| | Know all products of 1-digit numbers $\times 0$, $\times 3$, and $\times 9$ automatically. | 3, 5b 7a, 14, 17b |
| | Know all products of 1-digit numbers $\times 4$, $\times 6$, $\times 7$, and $\times 8$ automatically. | 3, 5b, 8, 17a–17e |
| | Divide within 100 fluently. | 3, 4 |
| 3.OA.8 | Assess the reasonableness of answers to problems. | 14, 22a, 22b |
| | Solve 2-step number stories involving two of the four operations. | 14 |
| | Model 2-step number stories with equations, using a letter or symbol for the unknown. | 14 |
| | Understand that grouping symbols affect the order in which operations are performed. | 5a, 5b, 7a, 7b |
| | Apply the order of operations when grouping symbols are not present. | 7a, 7b |
| 3.OA.9 | Identify arithmetic patterns and explain them using properties of operations. | 5a–5c |
| 3.NBT.2 | Add within 1,000 fluently. | 12, 20, 22a |
| | Subtract within 1,000 fluently. | 12, 19a, 22b |
| 3.NBT.3 | Multiply 1-digit numbers by multiples of 10. | 17a–17d, 19a |

# Goals (continued)

| CCSS Common Core State Standards | Goals for Mathematical Content (GMC) | Item(s) |
|---|---|---|
| **3.NF.1** | Understand, identify, and represent unit fractions as 1 part when a whole is divided into *b* equal parts. | 2, 13a |
| | Understand, identify, and represent non-unit fractions as the quantity formed by *a* parts of size $\frac{1}{b}$. | 6a, 13a |
| | Represent fractions by sharing collections of objects into equal shares. | 9 |
| **3.NF.2, 3.NF.2a** | Represent unit fractions on a number-line diagram. | 13a |
| **3.NF.2, 3.NF.2b** | Represent non-unit fractions on a number-line diagram. | 13a, 15 |
| **3.NF.3, 3.NF.3a** | Understand that equivalent fractions are the same size. | 10a–10c |
| | Understand that equivalent fractions name the same point on a number line. | 10a–10c, 13b |
| **3.NF.3, 3.NF.3b** | Recognize and generate simple equivalent fractions. | 10b |
| **3.NF.3, 3.NF.3c** | Express whole numbers as fractions. | 13a, 13b |
| | Recognize fractions that are equivalent to whole numbers. | 11a, 11b, 11e, 11f, 13a, 13b |
| **3.NF.3, 3.NF.3d** | Compare fractions with the same numerator or the same denominator. | 11c, 11d, 15 |
| | Recognize that fraction comparisons require the wholes to be the same size. | 6b, 10c |
| | Record fraction comparisons using >, =, or <. | 10b, 11a–11e, 13b |
| | Justify the conclusions of fraction comparisons. | 6b, 10a, 10c, 15 |
| **3.MD.1** | Measure time intervals in minutes. | 16 |
| | Solve number stories involving time intervals by adding or subtracting. | 16 |
| **3.MD.2** | Measure and estimate liquid volumes using liters and other units. | 12 |
| | Solve 1-step number stories involving volume. | 12 |
| **3.MD.4** | Measure lengths to the nearest $\frac{1}{2}$ inch, $\frac{1}{4}$ inch, or whole centimeter. | 1 |
| | Collect, organize, and represent data on line plots. | 1 |
| **3.MD.7, 3.MD.7b** | Multiply side lengths to find areas of rectangles. | 8, 19a |
| | Solve real-world and mathematical problems involving areas of rectangles. | 8 |
| | Represent whole-number products as rectangular areas. | 18 |
| **3.MD.7, 3.MD.7c** | Use area models to represent the distributive property. | 8 |
| **3.MD.7, 3.MD.7d** | Recognize area as additive. | 8, 19a |
| | Find areas of rectilinear figures by decomposing them into non-overlapping rectangles, and apply this technique to solve real-world problems. | 8, 19a |
| **3.MD.8** | Solve problems involving perimeters of polygons. | 18, 19a, 19b, 20 |
| | Exhibit rectangles with the same perimeter and different areas or the same area and different perimeters. | 18 |

# Goals (continued)

| CCSS Common Core State Standards | Goals for Mathematical Content (GMC) | Item(s) |
|---|---|---|
| 3.G.1 | Understand that shapes in different categories may share attributes that can define a larger category. | 21 |
| | Recognize specified subcategories of quadrilaterals. | 21 |
| | Draw quadrilaterals that do not belong to specified subcategories. | 21 |
| 3.G.2 | Partition shapes into parts with equal areas. | 6a |
| | Express the area of each part as a unit fraction of the whole. | 2 |

### Goals for Mathematical Practice (GMP)

| | | |
|---|---|---|
| SMP1 | Make sense of your problem.  **GMP1.1** | 4, 5a–5c, 8, 9, 10a–10c, 12, 14–16, 19a, 19b, 20 |
| | Check whether your answer makes sense.  **GMP1.4** | 14, 22a, 22b |
| SMP2 | Create mathematical representations using numbers, words, pictures, symbols, gestures, tables, graphs, and concrete objects.  **GMP2.1** | 6a, 12, 13a, 13b, 18 |
| | Make sense of the representations you and others use.  **GMP2.2** | 5a–5c, 6b, 7a, 7b, 10a–10c, 12, 19a, 19b |
| | Make connections between representations.  **GMP2.3** | 5a–5c, 10b |
| SMP3 | Make mathematical conjectures and arguments.  **GMP3.1** | 10a, 20 |
| | Make sense of others' mathematical thinking.  **GMP3.2** | 5a, 10a |
| SMP4 | Model real–world situations using graphs, drawings, tables, symbols, numbers, diagrams, and other representations.  **GMP4.1** | 1, 4, 14, 15 |
| | Use mathematical models to solve problems and answer questions.  **GMP4.2** | 4, 8, 15, 20 |
| SMP6 | Explain your mathematical thinking clearly and precisely.  **GMP6.1** | 3, 5c, 6b, 7b, 10a, 10c, 11f, 15, 16, 19b, 20, 21 |
| | Use clear labels, units, and mathematical language.  **GMP6.3** | 12, 14, 16, 19a, 20 |
| SMP7 | Look for mathematical structures such as categories, patterns, and properties.  **GMP7.1** | 5a, 21 |
| | Use structures to solve problems and answer questions.  **GMP7.2** | 3, 5b, 5c, 7a, 7b, 10a–10c, 17a–17e, 21 |
| SMP8 | Create and justify rules, shortcuts, and generalizations.  **GMP8.1** | 5 |

# End-of-Year Assessment

① Annabelle is measuring the widths of coins from around the world.
Create a line plot using her measurement data.

**Measures of World Coins**

| Measures of World Coins | |
|---|---|
| **Measure** | **Number of Coins** |
| $\frac{1}{2}$ in. | /// |
| $\frac{3}{4}$ in. | // |
| 1 in. | //// |
| $1\frac{1}{4}$ in. | // |
| $1\frac{1}{2}$ in. | / |

Width in Inches

$\frac{1}{2}$   $\frac{3}{4}$   1   $1\frac{1}{4}$   $1\frac{1}{2}$

Measure the widths of these two coins to the nearest $\frac{1}{4}$ inch.
Add the data to your line plot.

about _____ in.                about _____ in.

② Label each section of the fraction strip with a unit fraction.

# End-of-Year Assessment (continued)

③ Fill in the missing factor in the Fact Triangle.

Explain how you figured out the missing factor.

_____

_____

_____

_____

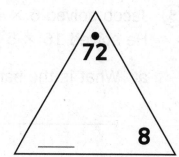

④ Five friends win $75 in a contest.
They agree to share the money equally.
How much money does each friend get?
The letter D represents the number of dollars per friend.

_____
  (number model with letter)

They have $10 bills and $1 bills.
Use numbers or pictures to show how much money each friend gets.

Answer: _____
   (unit)

# End-of-Year Assessment (continued)

⑤ Jacob solved 6 × 7 like this: (3 × 7) + (3 × 7) = 21 + 21 = 42.
He solved 16 × 5 like this: (8 × 5) + (8 × 5) = 40 + 40 = 80.

**a.** What is the same about Jacob's strategy for both problems?

_____

_____

**b.** Show how you can use Jacob's strategy to solve 8 × 9.

_____

**c.** Write another multiplication problem that you could solve
using Jacob's strategy.

_____

Explain how Jacob's strategy works for your problem.

_____

_____

⑥ **a.** Partition and shade the circles to show $\frac{2}{2} = \frac{6}{6}$.

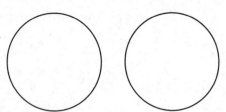

**b.** Explain why the circles shown above must be the same size.

_____

_____

# End-of-Year Assessment (continued)

## Rules for the Order of Operations

1.  Do operations inside parentheses first.
    Follow Rules 2 and 3 when computing inside parentheses.

2.  Then multiply or divide, in order, from left to right.

3.  Finally add or subtract, in order, from left to right.

⑦  **a.**  Use the order of operations to solve these number sentences.

$45 - 12 \times 0 =$ _____

$(45 - 12) \times 0 =$ _____

**b.**  Explain why the two number sentences have different answers.

_____

_____

_____

_____

⑧  Edie is planting a 6-foot by 15-foot flower garden.
    She is planting sunflowers in one part and daisies in the other part.

Daisies    Sunflowers

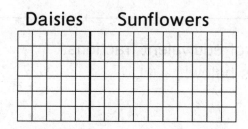

Write one or more number models that represent how you can find the area of the garden.

_____

Total area of Edie's garden: _____ square feet

# End-of-Year Assessment (continued)

⑨ A collection of 6 movie tickets is shared equally among 3 families.

How many tickets does each family get? _____ tickets

What fraction of the collection of movie tickets does each family get?

Each family gets $\dfrac{\phantom{0}}{\phantom{0}}$ of the tickets.

⑩ During a game of *Fraction Memory*, Marta turns over these two cards:

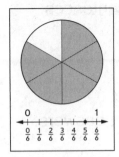 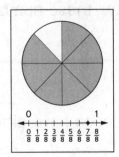

She thinks she found a pair of equivalent fractions.

**a.** Do you agree? Explain your thinking.

_____

_____

_____

_____

**b.** Use your fraction cards to find a pair of equivalent fractions.
Record your two fractions on the lines below.

_____ = _____

**c.** How do you know the fractions are equivalent?

_____

_____

_____

# End-of-Year Assessment (continued)

(11) Write >, <, or = to make the number sentences true.
You may use your fraction tools.

a. $1$ _____ $\frac{4}{4}$

b. $\frac{3}{3}$ _____ $\frac{4}{4}$

c. $\frac{3}{4}$ _____ $\frac{4}{4}$

d. $\frac{1}{2}$ _____ $\frac{1}{6}$

e. $\frac{3}{1}$ _____ $3$

f. What do you notice about the fractions in Parts a and b?

_____

_____

(12)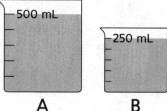

A          B

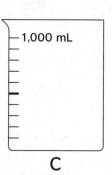

C

If you pour the water from Beakers A and B into
Beaker C, will Beaker C be full? *Hint:* 1 L = 1,000 mL _____

What volume of water will be in Beaker C after
you pour in water from Beakers A and B? about _____
(unit)

Shade Beaker C to show the total liquid volume.
How much more water would you need to fill Beaker C?

about _____
(unit)

# End-of-Year Assessment (continued)

**13**  **a.** Partition this number line into eighths. Label with fractions.

0                                                                    1

**b.** Compare these fractions. Write >, <, or = to make the number sentences true. Use your number line.

$\frac{8}{8}$ _____ 1                    $\frac{2}{8}$ _____ $\frac{1}{2}$

**14** Arjun has 12 eggs.
He uses 2 eggs for each omelet and makes 3 omelets.
How many eggs does he have left?

Write one or more number models that match the story.
Use a letter for what you are trying to find out.

The letter _____ represents _____.

_____
(number model(s) with letter)

Arjun has _____ left.
                           (unit)
Check whether your answer makes your number model(s) true.
Write your number model(s) with your answer.

_____

## End-of-Year Assessment (continued)

⑮ Sylvie ran $\frac{3}{8}$ of a mile. Ivan ran $\frac{3}{4}$ of a mile.
Partition and use the number lines below to show how far they ran.

### Sylvie

0                                   1

### Ivan

0                                   1

Who ran farther? Explain how you know.

_____

_____

_____

⑯ Mario's baseball practice ends at 7:30 P.M.
His mom leaves to pick him up at 7:15 P.M.
It takes her 25 minutes to get to the baseball field.

Will she arrive on time? Explain.

_____

_____

# End-of-Year Assessment (continued)

**17** Solve the extended multiplication facts.

**a.** 60 × 7 = _____     **b.** _____ = 40 × 3

**c.** _____ = 70 × 8     **d.** 6 × _____ = 240

**e.** What basic fact could help you solve Part d?

_____

**18** Draw two different rectangles that each have an area of 36 square units. Label your rectangles A and B. Write a number model for finding the area of each rectangle.

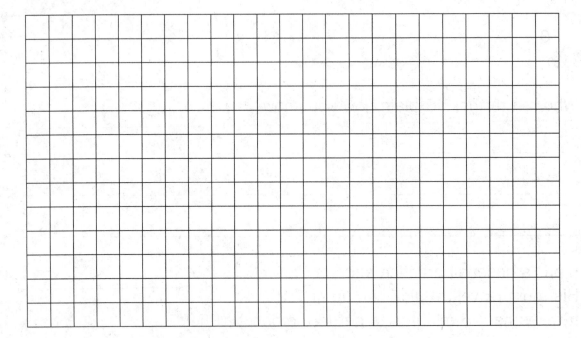

The perimeter of Rectangle A is _____ units.

The perimeter of Rectangle B is _____ units.

# End-of-Year Assessment (continued)

**19** **a.** Find the missing side lengths of this rectilinear figure. Then find the area. Remember to include the unit.

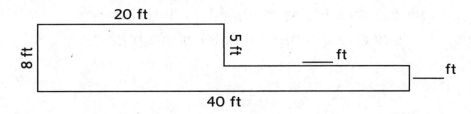

The letter *A* represents the total area of the figure.

Number model(s): _____

Area: _____

(unit)

**b.** How did you figure out the missing side lengths?

_____

_____

_____

_____

_____

# End-of-Year Assessment (continued)

20 Two third-grade teams run races at Field Day. They run around the rectangular fields marked with cones and compare times.

Gabriel says that the race is not fair because the distance around Field 1 is longer. Find the perimeter of each field.

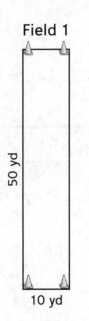

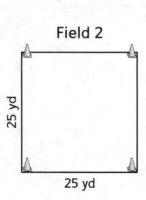

Field 1

50 yd

10 yd

Field 2

25 yd

25 yd

Perimeter of Field 1: _____
                                              (unit)

Perimeter of Field 2: _____
                                              (unit)

Is the race fair? Explain your answer.

_____

_____

_____

# End-of-Year Assessment (continued)

**21** Circle all the rectangles, mark an X on all the squares, and shade all the rhombuses.

  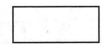   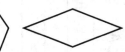

Explain why the shapes you circled are rectangles.

_____

_____

Draw another quadrilateral that is NOT a rectangle, a square, or a rhombus.

**22** Solve. Make an estimate to check whether your answer makes sense.

**Unit**

**a.** Estimate:

_____

```
  4 6 1
+ 2 6 9
```

**b.** Estimate:

_____

```
  3 4 8
- 1 5 4
```

# Multiplication Facts Assessment Tool Page 1

Key: F = fluent; A = automatic

Set 1: 2s, 5s, and 10s Facts

| Name | 2 × 2 | 2 × 3 | 2 × 4 | 2 × 5 | 2 × 6 | 2 × 7 | 2 × 8 | 2 × 9 | 2 × 10 | NOTES |
|------|-------|-------|-------|-------|-------|-------|-------|-------|--------|-------|
| 1. | | | | | | | | | | |
| 2. | | | | | | | | | | |
| 3. | | | | | | | | | | |
| 4. | | | | | | | | | | |
| 5. | | | | | | | | | | |
| 6. | | | | | | | | | | |
| 7. | | | | | | | | | | |
| 8. | | | | | | | | | | |
| 9. | | | | | | | | | | |
| 10. | | | | | | | | | | |
| 11. | | | | | | | | | | |
| 12. | | | | | | | | | | |
| 13. | | | | | | | | | | |
| 14. | | | | | | | | | | |
| 15. | | | | | | | | | | |
| 16. | | | | | | | | | | |
| 17. | | | | | | | | | | |
| 18. | | | | | | | | | | |
| 19. | | | | | | | | | | |
| 20. | | | | | | | | | | |
| 21. | | | | | | | | | | |
| 22. | | | | | | | | | | |
| 23. | | | | | | | | | | |
| 24. | | | | | | | | | | |
| 25. | | | | | | | | | | |

# Multiplication Facts Assessment Tool Page 2

Key: F = fluent; A = automatic

Set 1: 2s, 5s, and 10s Facts

| Name | 5 × 3 | 5 × 4 | 5 × 5 | 5 × 6 | 5 × 7 | 5 × 8 | 5 × 9 | 5 × 10 | NOTES |
|---|---|---|---|---|---|---|---|---|---|
| 1. | | | | | | | | | |
| 2. | | | | | | | | | |
| 3. | | | | | | | | | |
| 4. | | | | | | | | | |
| 5. | | | | | | | | | |
| 6. | | | | | | | | | |
| 7. | | | | | | | | | |
| 8. | | | | | | | | | |
| 9. | | | | | | | | | |
| 10. | | | | | | | | | |
| 11. | | | | | | | | | |
| 12. | | | | | | | | | |
| 13. | | | | | | | | | |
| 14. | | | | | | | | | |
| 15. | | | | | | | | | |
| 16. | | | | | | | | | |
| 17. | | | | | | | | | |
| 18. | | | | | | | | | |
| 19. | | | | | | | | | |
| 20. | | | | | | | | | |
| 21. | | | | | | | | | |
| 22. | | | | | | | | | |
| 23. | | | | | | | | | |
| 24. | | | | | | | | | |
| 25. | | | | | | | | | |

# Multiplication Facts Assessment Tool Page 3

Set 1: 2s, 5s, and 10s Facts

Key: F = fluent; A = automatic

| Name | 10 × 3 | 10 × 4 | 10 × 6 | 10 × 7 | 10 × 8 | 10 × 9 | 10 × 10 | NOTES |
|------|--------|--------|--------|--------|--------|--------|---------|-------|
| 1. | | | | | | | | |
| 2. | | | | | | | | |
| 3. | | | | | | | | |
| 4. | | | | | | | | |
| 5. | | | | | | | | |
| 6. | | | | | | | | |
| 7. | | | | | | | | |
| 8. | | | | | | | | |
| 9. | | | | | | | | |
| 10. | | | | | | | | |
| 11. | | | | | | | | |
| 12. | | | | | | | | |
| 13. | | | | | | | | |
| 14. | | | | | | | | |
| 15. | | | | | | | | |
| 16. | | | | | | | | |
| 17. | | | | | | | | |
| 18. | | | | | | | | |
| 19. | | | | | | | | |
| 20. | | | | | | | | |
| 21. | | | | | | | | |
| 22. | | | | | | | | |
| 23. | | | | | | | | |
| 24. | | | | | | | | |
| 25. | | | | | | | | |

# Multiplication Facts Assessment Tool Page 4

Key: F = fluent; A = automatic

Set 2: Remaining Squares

| Name | 3 × 3 | 4 × 4 | 6 × 6 | 7 × 7 | 8 × 8 | 9 × 9 | NOTES |
|------|-------|-------|-------|-------|-------|-------|-------|
| 1. | | | | | | | |
| 2. | | | | | | | |
| 3. | | | | | | | |
| 4. | | | | | | | |
| 5. | | | | | | | |
| 6. | | | | | | | |
| 7. | | | | | | | |
| 8. | | | | | | | |
| 9. | | | | | | | |
| 10. | | | | | | | |
| 11. | | | | | | | |
| 12. | | | | | | | |
| 13. | | | | | | | |
| 14. | | | | | | | |
| 15. | | | | | | | |
| 16. | | | | | | | |
| 17. | | | | | | | |
| 18. | | | | | | | |
| 19. | | | | | | | |
| 20. | | | | | | | |
| 21. | | | | | | | |
| 22. | | | | | | | |
| 23. | | | | | | | |
| 24. | | | | | | | |
| 25. | | | | | | | |

# Multiplication Facts Assessment Tool Page 5

Set 3: 3s and 9s Facts    Key: F = fluent; A = automatic

| Name | 3 × 4 | 3 × 6 | 3 × 7 | 3 × 8 | 9 × 9 | 9 × 4 | 9 × 6 | 9 × 7 | 9 × 8 | NOTES |
|------|-------|-------|-------|-------|-------|-------|-------|-------|-------|-------|
| 1. | | | | | | | | | | |
| 2. | | | | | | | | | | |
| 3. | | | | | | | | | | |
| 4. | | | | | | | | | | |
| 5. | | | | | | | | | | |
| 6. | | | | | | | | | | |
| 7. | | | | | | | | | | |
| 8. | | | | | | | | | | |
| 9. | | | | | | | | | | |
| 10. | | | | | | | | | | |
| 11. | | | | | | | | | | |
| 12. | | | | | | | | | | |
| 13. | | | | | | | | | | |
| 14. | | | | | | | | | | |
| 15. | | | | | | | | | | |
| 16. | | | | | | | | | | |
| 17. | | | | | | | | | | |
| 18. | | | | | | | | | | |
| 19. | | | | | | | | | | |
| 20. | | | | | | | | | | |
| 21. | | | | | | | | | | |
| 22. | | | | | | | | | | |
| 23. | | | | | | | | | | |
| 24. | | | | | | | | | | |
| 25. | | | | | | | | | | |

# Multiplication Facts Assessment Tool Page 6

Key: F = fluent; A = automatic

Set 4: Remaining Facts

| Name | 4 × 6 | 4 × 7 | 4 × 8 | 6 × 7 | 6 × 8 | 7 × 8 | NOTES |
|------|-------|-------|-------|-------|-------|-------|-------|
| 1. | | | | | | | |
| 2. | | | | | | | |
| 3. | | | | | | | |
| 4. | | | | | | | |
| 5. | | | | | | | |
| 6. | | | | | | | |
| 7. | | | | | | | |
| 8. | | | | | | | |
| 9. | | | | | | | |
| 10. | | | | | | | |
| 11. | | | | | | | |
| 12. | | | | | | | |
| 13. | | | | | | | |
| 14. | | | | | | | |
| 15. | | | | | | | |
| 16. | | | | | | | |
| 17. | | | | | | | |
| 18. | | | | | | | |
| 19. | | | | | | | |
| 20. | | | | | | | |
| 21. | | | | | | | |
| 22. | | | | | | | |
| 23. | | | | | | | |
| 24. | | | | | | | |
| 25. | | | | | | | |

# Multiplication Facts Assessment Tool Page 7

Key: F = fluent; A = automatic

× 0, × 1, and Turn-Around Rule

| Name | Understands × 0 | Understands × 1 | Understands Turn-Around Rule | OTHER NOTES |
|------|-----------------|-----------------|------------------------------|-------------|
| 1. | | | | |
| 2. | | | | |
| 3. | | | | |
| 4. | | | | |
| 5. | | | | |
| 6. | | | | |
| 7. | | | | |
| 8. | | | | |
| 9. | | | | |
| 10. | | | | |
| 11. | | | | |
| 12. | | | | |
| 13. | | | | |
| 14. | | | | |
| 15. | | | | |
| 16. | | | | |
| 17. | | | | |
| 18. | | | | |
| 19. | | | | |
| 20. | | | | |
| 21. | | | | |
| 22. | | | | |
| 23. | | | | |
| 24. | | | | |
| 25. | | | | |

# Unit ____ Assessment Check-Ins
## Individual Profile of Progress

| Lesson | CCSS | Assess Progress | Comments |
|--------|------|-----------------|----------|
|        |      |                 |          |
|        |      |                 |          |
|        |      |                 |          |
|        |      |                 |          |
|        |      |                 |          |
|        |      |                 |          |
|        |      |                 |          |
|        |      |                 |          |
|        |      |                 |          |
|        |      |                 |          |
|        |      |                 |          |

**Assess Progress**   **M** = meeting expectations   **N** = not meeting expectations   **N/A** = not assessed

# Unit ____ Assessment Check-Ins: Class Checklist

| | Assessment Check-In Items by Lesson | | | | | | | | | | |
|---|---|---|---|---|---|---|---|---|---|---|---|
| **Names** | | | | | | | | | | | |
| | | | | | | | | | | | |
| | | | | | | | | | | | |
| | | | | | | | | | | | |
| | | | | | | | | | | | |
| | | | | | | | | | | | |
| | | | | | | | | | | | |
| | | | | | | | | | | | |
| | | | | | | | | | | | |
| | | | | | | | | | | | |
| | | | | | | | | | | | |
| | | | | | | | | | | | |
| | | | | | | | | | | | |
| | | | | | | | | | | | |
| | | | | | | | | | | | |
| | | | | | | | | | | | |
| | | | | | | | | | | | |
| | | | | | | | | | | | |
| | | | | | | | | | | | |
| | | | | | | | | | | | |
| | | | | | | | | | | | |
| | | | | | | | | | | | |
| | | | | | | | | | | | |
| | | | | | | | | | | | |
| | | | | | | | | | | | |
| | | | | | | | | | | | |
| | | | | | | | | | | | |
| | | | | | | | | | | | |

**Assess Progress**    **M** = meeting expectations    **N** = not meeting expectations    **N/A** = not assessed

# Unit _____ Progress Check
## Individual Profile of Progress

| Item(s) | CCSS | Assess Progress | Comments |
|---------|------|-----------------|----------|
| **Unit Assessment** | | | |
| | | | |
| | | | |
| | | | |
| | | | |
| | | | |
| | | | |
| | | | |
| | | | |
| | | | |
| | | | |
| | | | |
| | | | |
| | | | |
| **Open Response Assessment (odd-numbered units)** | | | |
| | | | |

**Assess Progress**     **M** = meeting expectations    **N** = not meeting expectations    **N/A** = not assessed

NAME _____ DATE _____

# Unit ____ **Progress Check**
## Individual Profile of Progress

| Item(s) | CCSS | Assess Progress | Comments |
|---|---|---|---|
| **Challenge (Optional)** | | | |
| | | | |

**Assess Progress**     **M** = meeting expectations     **N** = not meeting expectations     **N/A** = not assessed

---

✂ - - - - - - - - - - - - - - - - - - - - - - - - - - - - - - - - - - - -

NAME _____ DATE _____

# Unit ____ **Progress Check**
## Individual Profile of Progress

| Item(s) | CCSS | Assess Progress | Comments |
|---|---|---|---|
| **Challenge (Optional)** | | | |
| | | | |

**Assess Progress**     **M** = meeting expectations     **N** = not meeting expectations     **N/A** = not assessed

# Unit ____ Progress Check: Class Checklist

| | Unit Assessment Items | | | | | | | | Challenge (Optional) | Open Response Assessment |
|---|---|---|---|---|---|---|---|---|---|---|
| **Names** | | | | | | | | | | |
| | | | | | | | | | | |
| | | | | | | | | | | |
| | | | | | | | | | | |
| | | | | | | | | | | |
| | | | | | | | | | | |
| | | | | | | | | | | |
| | | | | | | | | | | |
| | | | | | | | | | | |
| | | | | | | | | | | |
| | | | | | | | | | | |
| | | | | | | | | | | |
| | | | | | | | | | | |
| | | | | | | | | | | |
| | | | | | | | | | | |
| | | | | | | | | | | |
| | | | | | | | | | | |
| | | | | | | | | | | |
| | | | | | | | | | | |
| | | | | | | | | | | |
| | | | | | | | | | | |
| | | | | | | | | | | |
| | | | | | | | | | | |
| | | | | | | | | | | |
| | | | | | | | | | | |
| | | | | | | | | | | |
| | | | | | | | | | | |
| | | | | | | | | | | |

**Assess Progress**      **M** = meeting expectations      **N** = not meeting expectations      **N/A** = not assessed

# Unit ____ Progress Check
## Individual Profile of Progress

| Item(s) | CCSS | Assess Progress | Comments |
|---------|------|-----------------|----------|
| **Cumulative Assessment** | | | |
| | | | |
| | | | |
| | | | |
| | | | |
| | | | |
| | | | |
| | | | |
| | | | |
| | | | |
| | | | |
| | | | |

**Assess Progress**        **M** = meeting expectations    **N** = not meeting expectations    **N/A** = not assessed

# Unit ___ Progress Check: Class Checklist

| | Cumulative Assessment Items | | | | | | | | | |
|---|---|---|---|---|---|---|---|---|---|---|
| **Names** | | | | | | | | | | |
| | | | | | | | | | | |
| | | | | | | | | | | |
| | | | | | | | | | | |
| | | | | | | | | | | |
| | | | | | | | | | | |
| | | | | | | | | | | |
| | | | | | | | | | | |
| | | | | | | | | | | |
| | | | | | | | | | | |
| | | | | | | | | | | |
| | | | | | | | | | | |
| | | | | | | | | | | |
| | | | | | | | | | | |
| | | | | | | | | | | |
| | | | | | | | | | | |
| | | | | | | | | | | |
| | | | | | | | | | | |
| | | | | | | | | | | |
| | | | | | | | | | | |
| | | | | | | | | | | |
| | | | | | | | | | | |
| | | | | | | | | | | |
| | | | | | | | | | | |
| | | | | | | | | | | |
| | | | | | | | | | | |
| | | | | | | | | | | |
| | | | | | | | | | | |
| | | | | | | | | | | |

**Assess Progress**     **M** = meeting expectations     **N** = not meeting expectations     **N/A** = not assessed

# Mathematical Practices for Unit(s) _____
## Individual Profile of Progress

| Use this sheet to record children's use of the mathematical practices in lesson activities, Assessment Check-Ins, Writing/Reasoning prompts, and Progress Checks. | Opportunity | Date | Comments |
|---|---|---|---|
| **SMP1: Make sense of problems and persevere in solving them.** | | | |
| **GMP1.1** Make sense of your problem. | | | |
| **GMP1.2** Reflect on your thinking as you solve your problem. | | | |
| **GMP1.3** Keep trying when your problem is hard. | | | |
| **GMP1.4** Check whether your answer makes sense. | | | |
| **GMP1.5** Solve problems in more than one way. | | | |
| **GMP1.6** Compare the strategies you and others use. | | | |
| **SMP2: Reason abstractly and quantitatively.** | | | |
| **GMP2.1** Create mathematical representations using numbers, words, pictures, symbols, gestures, tables, graphs, and concrete objects. | | | |
| **GMP2.2** Make sense of the representations you and others use. | | | |
| **GMP2.3** Make connections between representations. | | | |
| **SMP3: Construct viable arguments and critique the reasoning of others.** | | | |
| **GMP3.1** Make mathematical conjectures and arguments. | | | |
| **GMP3.2** Make sense of others' mathematical thinking. | | | |
| **SMP4: Model with mathematics.** | | | |
| **GMP4.1** Model real-world situations using graphs, drawings, tables, symbols, numbers. diagrams, and other representations. | | | |
| **GMP4.2** Use mathematical models to solve problems and answer questions. | | | |

# Mathematical Practices
## Individual Profile of Progress (continued)

| | Opportunity | Date | Comments |
|---|---|---|---|
| **SMP5: Use appropriate tools strategically.** | | | |
| **GMP5.1** Choose appropriate tools. | | | |
| **GMP5.2** Use tools effectively and make sense of your results. | | | |
| | | | |
| | | | |
| | | | |
| **SMP6: Attend to precision.** | | | |
| **GMP6.1** Explain your mathematical thinking clearly and precisely. | | | |
| **GMP6.2** Use an appropriate level of precision for your problem. | | | |
| **GMP6.3** Use clear labels, units, and mathematical language. | | | |
| **GMP6.4** Think about accuracy and efficiency when you count, measure, and calculate. | | | |
| **SMP7: Look for and make use of structure.** | | | |
| **GMP7.1** Look for mathematical structures such as categories, patterns, and properties. | | | |
| **GMP7.2** Use structures to solve problems and answer questions. | | | |
| | | | |
| | | | |
| **SMP8: Look for and express regularity in repeated reasoning.** | | | |
| **GMP8.1** Create and justify rules, shortcuts, and generalizations. | | | |
| | | | |
| | | | |
| | | | |
| | | | |

# Mathematical Practice Opportunities
## Class Record

Standard for Mathematical Practice: _____

Goal for Mathematical Practice: _____

Opportunity: _____

| Names | +/✓/− | Comments |
|-------|-------|----------|
|       |       |          |
|       |       |          |
|       |       |          |
|       |       |          |
|       |       |          |
|       |       |          |
|       |       |          |
|       |       |          |
|       |       |          |
|       |       |          |
|       |       |          |
|       |       |          |
|       |       |          |
|       |       |          |
|       |       |          |
|       |       |          |
|       |       |          |
|       |       |          |
|       |       |          |
|       |       |          |
|       |       |          |
|       |       |          |
|       |       |          |
|       |       |          |
|       |       |          |

# Mathematical Practices in Open Response Problems
## Individual Profile of Progress

| Lesson | SMP | GMP | Assess Progress | Comments |
|--------|-----|-----|-----------------|----------|
| **Open Response and Reengagement** | | | | |
| 1-6 | 4 | 4.2 | | |
| 2-8 | 2 | 2.1 | | |
| 3-2 | 6 | 6.1 | | |
| 4-11 | 4 | 4.1 | | |
| 5-10 | 1 | 1.1 | | |
| 6-9 | 7 | 7.1 | | |
| 7-8 | 8 | 8.1 | | |
| 8-4 | 3 | 3.1 | | |
| 9-6 | 5 | 5.2 | | |
| **Progress Check** | | | | |
| 1-14 | 4 | 4.2 | | |
| 3-14 | 3 | 3.2 | | |
| 5-12 | 6 | 6.4 | | |
| 7-13 | 3 | 3.1 | | |
| 9-8 | 7 | 7.1 | | |

**Assess Progress**  **E** = exceeding expectations  **M** = meeting expectations
**P** = partially meeting expectations  **N** = not meeting expectations

# Mathematical Practices in Open Response Problems
## Class Checklist

| Names | Open Response and Reengagement Items | | | | | | | | | Progress Check Items | | | | |
|---|---|---|---|---|---|---|---|---|---|---|---|---|---|---|
| | SMP4 GMP4.2 | SMP2 GMP2.1 | SMP6 GMP6.1 | SMP4 GMP4.1 | SMP1 GMP1.1 | SMP7 GMP7.1 | SMP8 GMP8.1 | SMP3 GMP3.1 | SMP5 GMP5.2 | SMP4 GMP4.2 | SMP3 GMP3.2 | SMP6 GMP6.4 | SMP3 GMP3.1 | SMP7 GMP7.1 |
| | 1-6 | 2-8 | 3-2 | 4-11 | 5-10 | 6-9 | 7-8 | 8-4 | 9-6 | 1-14 | 3-14 | 5-12 | 7-13 | 9-8 |
| | | | | | | | | | | | | | | |
| | | | | | | | | | | | | | | |
| | | | | | | | | | | | | | | |
| | | | | | | | | | | | | | | |
| | | | | | | | | | | | | | | |
| | | | | | | | | | | | | | | |
| | | | | | | | | | | | | | | |
| | | | | | | | | | | | | | | |
| | | | | | | | | | | | | | | |
| | | | | | | | | | | | | | | |
| | | | | | | | | | | | | | | |
| | | | | | | | | | | | | | | |
| | | | | | | | | | | | | | | |
| | | | | | | | | | | | | | | |
| | | | | | | | | | | | | | | |
| | | | | | | | | | | | | | | |
| | | | | | | | | | | | | | | |
| | | | | | | | | | | | | | | |
| | | | | | | | | | | | | | | |
| | | | | | | | | | | | | | | |
| | | | | | | | | | | | | | | |
| | | | | | | | | | | | | | | |
| | | | | | | | | | | | | | | |
| | | | | | | | | | | | | | | |
| | | | | | | | | | | | | | | |

**Assess Progress**  E = exceeding expectations  M = meeting expectations
P = partially meeting expectations  N = not meeting expectations

# Parent Reflections

**Use some of the following questions (or your own) and tell us how you see your child progressing in mathematics.**

*Do you see evidence of your child using mathematics at home?*

*What do you think are your child's strengths and challenges in mathematics?*

*Does your child demonstrate responsibility for completing Home Links?*

*What thoughts do you have about your child's progress in mathematics?*

| NAME | DATE | TIME |
|------|------|------|

# My Exit Slip

| NAME | DATE | TIME |
|------|------|------|

# My Exit Slip

# About My Math Class A

Draw a face or write the words
that show how you feel.

Good          OK          Not so good

| | |
|---|---|
| ① This is how I feel about math: | ② This is how I feel about working with a partner or in a group: | ③ This is how I feel about working by myself: |
| ④ This is how I feel about solving number stories: | ⑤ This is how I feel about doing Home Links with my family: | ⑥ This is how I feel about finding new ways to solve problems: |

Circle **yes**, **sometimes**, or **no**.

⑦ I like to figure things out. I am curious.

**yes**        **sometimes**        **no**

⑧ I keep trying even when I don't understand something right away.

**yes**        **sometimes**        **no**

# About My Math Class B

Circle the word that best describes how you feel.

1. I enjoy mathematics class. **yes** **sometimes** **no**

2. I like to work with a partner or in a group. **yes** **sometimes** **no**

3. I like to work by myself. **yes** **sometimes** **no**

4. I like to solve problems in mathematics. **yes** **sometimes** **no**

5. I enjoy doing Home Links with my family. **yes** **sometimes** **no**

6. In mathematics, I am good at _____

   _____

   _____

   _____

7. One thing I like about mathematics is _____

   _____

   _____

8. One thing I find difficult in mathematics is _____

   _____

   _____

| NAME | DATE | TIME |
|------|------|------|

# Math Log A

What did you learn in mathematics this week?

_____

_____

_____

_____

_____

✂ - - - - - - - - - - - - - - - - - - - - - - - - - - - - - - - - - - -

| NAME | DATE | TIME |
|------|------|------|

# Math Log A

What did you learn in mathematics this week?

_____

_____

_____

_____

_____

# Math Log B

Question: _____

_____

_____

_____

_____

_____

# Math Log B

Question: _____

_____

_____

_____

_____

_____

NAME         DATE         TIME

# Math Log C

Work Box

Tell how you solved this problem.

_____

_____

_____

_____

_____

_____

_____

NAME         DATE         TIME

# Math Log C

Work Box

Tell how you solved this problem.

_____

_____

_____

_____

_____

_____

_____

| NAME | DATE | TIME |  |

# Good Work!

 I have chosen this work because _____

_____

_____

_____

_____

_____

_____

✂ - - - - - - - - - - - - - - - - - - - - - - - - - - - - -

| NAME | DATE | TIME |  |

# Good Work!

 I have chosen this work because _____

_____

_____

_____

_____

_____

_____

NAME                    DATE                    TIME

# My Work

This work shows that I can _____

_____

_____

I am still learning to _____

_____

_____

_____

✂ - - - - - - - - - - - - - - - - - - - - - - - - - - - - - - - - - - -

NAME                    DATE                    TIME

# My Work

This work shows that I can _____

_____

_____

I am still learning to _____

_____

_____

_____

# Name-Collection Boxes

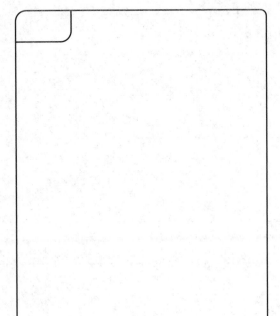

①

②

③

④

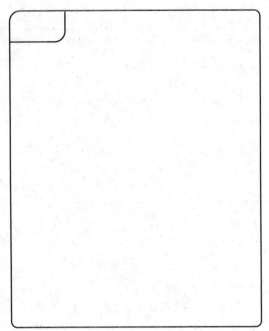

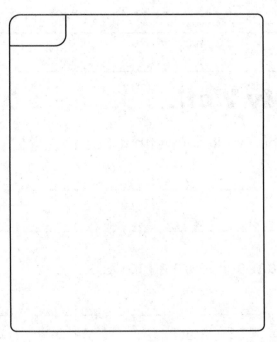

# Interim Assessment Answers

## Beginning-of-Year Assessment

**1. a.** 7:00      **b.** 11:15

**2. a.**       **b.**

**3. a.** 100      **b.** 46

**4.** Sample answer: $28 + ? = 57$; $57 - 28 = ?$

Sample answer:

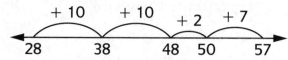

I counted up by 10s from 28 to 48, then 2 more to get to 50, and then 7 more to get to 57. That is 29 in all.

29

**5.**

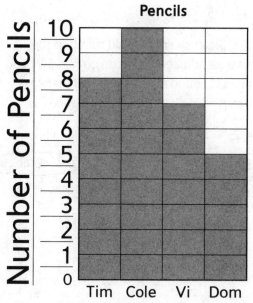

**a.** Cole    **b.** Dom    **c.** 5      **d.** 30

**6.** Sample rules given.
Rule: $- 10$, count back by 10s, subtract 10

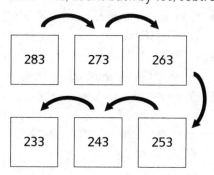

**7.** Sample answers: $30 + 5 = ?$, $35 - 18 = ?$; $30 + 5 - 18 = ?$

17

**8. a.** 15

**b.** Sample answers: $5 + 5 + 5 = 15$; $3 \times 5 = 15$

**9. a.** 6; 4; 8; 14; 9; 10

**b.** $6 = 3 + 3$    $5 + 6 = 11$    $20 - 10 = 10$

**10.**

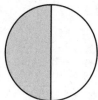

Sample answers: one out of two equal parts; one-half; 1-half

## Mid-Year Assessment

**1. a.** Sample answer: The array shows 2 rows with 9 in each row. There are 2 groups of 9, and $9 + 9 = 18$.

**b.**

Sample answer: I added one row to make 3 groups of 9. So $3 \times 9 = 18 + 9 = 27$.

27

**2.** 5 inches

Sample answers: $20 \div 4 = 5$; $4 \times 5 = 20$

**3.** Drawings vary.

6 stickers

Sample answers: $30 \div 5 = 6$; $5 \times 6 = 30$

**4. a.** Rule: $\times 2$ or double

| in | out |
|----|-----|
| 10 | 20 |
| 7  | 14 |
| 3  | 6  |
| 6  | 12 |
| Answers | vary. |

**b.** Sample answer: 7 is doubled to get 14, and 3 is doubled to get 6. So the rule is multiply by 2.

**5.**

| in | out |
|----|-----|
| 3  | 30  |
| 10 | 100 |
| 7  | 70  |
| 2  | 20  |
| Answers | vary. |

**6. a.** Sample number models: $4 \times 4 = 16$; $16 - 3 = 13$

Representations vary.

13 balloons

**b.** Sample answer: It makes sense because 3 balloons popped, so there are 3 fewer balloons than 16.

**7. a.** Sample answer: $600 + 300 = 900$

**b.** No. Sample explanation: Her answer is 817, and that's only 17 more than 800 instead of $43 + 91$ more. She should have added $40 + 90$ to get 130 instead of 13.

**c.** 934

**8.** $363 - 318 = ?$ or $318 + ? = 363$

Representations vary.

45

**9. a.** $690 + 230 = 920$

**b.** $700 + 200 = 900$

**c.** Sample answers: I would round to the nearest 100 because I can do it in my head. I would round to the nearest 10 because that sum would be closer to 1,000.

**10. a.** 4:05   **b.** 4:35   **c.** 30 minutes

**11.** paper clip    pineapple    centimeter cube

dime    bottle of water    baseball bat

## Mid-Year Assessment (continued)

**12. a.**

6 cm

4 cm

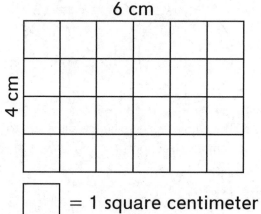

□ = 1 square centimeter

24

**b.** 4 × 6 = 24 or 6 × 4 = 24

**13. a.**

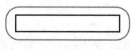

**b.** Sample answers: They have 4 sides and 4 right angles. I circled the square because it is a special rectangle that has 4 equal-length sides.

**14. a.**

### Number of Blankets

| | |
|---|---|
| small | □ □ 𝈒 |
| medium | □ □ □ □ □ □ □ |
| large | □ □ □ □ □ 𝈒 |
| extra large | □ □ |

Key: □ = 20 blankets

**b.** 50 + 160 + 110 = ?

320

## End-of-Year Assessment

**1.** **Measures of World Coins**

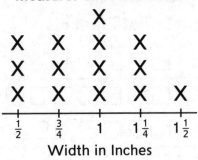

Width in Inches

$\frac{3}{4}$; $1\frac{1}{4}$

**2.**

| $\frac{1}{6}$ | $\frac{1}{6}$ | $\frac{1}{6}$ | $\frac{1}{6}$ | $\frac{1}{6}$ | $\frac{1}{6}$ |
|---|---|---|---|---|---|

**3.** 9; Sample answers: I thought *8 times what number is 72?* I started with 80 ÷ 8 = 10, subtracted one group of 8, and got 72. So 9 groups of 8 is 72.

**4.** $75 ÷ 5 = D$ or $5 \times D = \$75$

Sample answer:

$10  
$1 $1 $1  
$1 $1

$10  
$1 $1 $1  
$1 $1

$10  
$1 $1 $1  
$1 $1

$10  
$1 $1 $1  
$1 $1

$10  
$1 $1 $1  
$1 $1

$15

**5.** **a.** Sample answers: Jacob used doubling. He broke apart the even factors.

**b.** $(4 \times 9) + (4 \times 9) = 72$

**c.** Sample answer: $3 \times 12 = ?$

Sample answer: I can break 12 into 6 and 6. Then I can multiply $3 \times 6$ twice and get $18 + 18 = 36$, so $3 \times 12 = 36$.

**6.** **a.**

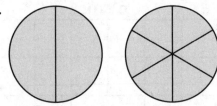

**b.** The wholes have to be the same size to compare fractions.

**7.** **a.** 45; 0

**b.** Sample answer: In the first problem, $12 \times 0$ is solved first. In the second problem, $45 - 12$ is solved first.

**8.** Sample answer: $(6 \times 5) + (6 \times 10) = ?$

90

**9.** 2; $\frac{2}{6}$ or $\frac{1}{3}$

**10.** **a.** No. Sample explanations: The shaded area of each circle on the cards is not the same size. On the number line, $\frac{7}{8}$ is farther from 0 than $\frac{5}{6}$ is. There is a smaller piece missing from $\frac{7}{8}$ than from $\frac{5}{6}$, so $\frac{7}{8}$ is larger than $\frac{5}{6}$.

**b.** Answers vary.

**c.** Sample answers: The shaded area of each circle is the same. The points on the number lines are the same distance from 0.

**11.** **a.** =    **b.** =    **c.** <    **d.** >    **e.** =

**f.** Sample answers: They are all equal. They all equal 1.

**12.** No; 750 mL

250 mL

## End-of-Year Assessment (continued)

**13. a.** Sample answer:

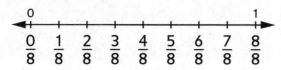

**b.** =, <

**14.** Sample answers: $E$; the number of eggs left over

Sample answers: $2 \times 3 = 6$ and $12 - 6 = E$; $12 - (2 \times 3) = E$; $2 \times 3 + E = 12$

6 eggs

Sample answers: $2 \times 3 = 6$ and $12 - 6 = 6$; $12 - (2 \times 3) = 6$; $2 \times 3 + 6 = 12$

**15.**

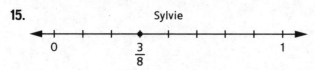

Ivan ran farther. Sample explanation: $\frac{3}{4}$ is a greater distance from 0 than $\frac{3}{8}$ is, so Ivan ran farther than Sylvie.

**16.** No. Sample explanation: 25 minutes after 7:15 is 7:40. She will be 10 minutes late.

**17. a.** 420    **b.** 120    **c.** 560    **d.** 40

**e.** $6 \times 4 = 24$

**18.** Sample answers given.

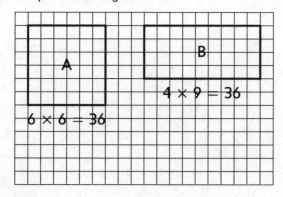

24; 26

**19. a.**

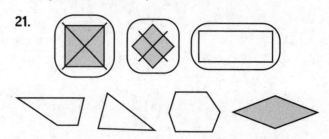

Sample answers: $(20 \times 8) + (20 \times 3) = A$

220 square feet

**b.** Sample answer: I know that opposite sides of the rectangles have to be the same length. I know that the length of one side is 40 feet. Because part of the opposite side is 20 feet, the other part is also 20 feet. I know that one side is 8 feet. Because part of the opposite side is 5 feet, the other part is 3 feet.

**20.** 120 yards; 100 yards

No. Sample explanation: The perimeter of Field 1 is greater than the perimeter of Field 2.

**21.**

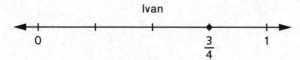

Sample answer: They all have parallel, equal-length opposite sides. They have four right angles.

Sample answer:

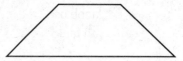

**22.** Sample estimates given.

**a.** $450 + 250 = 700$

730

**b.** $350 - 150 = 200$

194

# GMP1.1 Rubric

## SMP1: Make sense of problems and persevere in solving them.

| Goal for Mathematical Practice | Not Meeting Expectations | Partially Meeting Expectations | Meeting Expectations | Exceeding Expectations |
|---|---|---|---|---|
| **GMP1.1** Make sense of your problem. | | | | |

Generic GMP Rubrics

# GMP1.2 Rubric

## SMP1: Make sense of problems and persevere in solving them.

| Goal for Mathematical Practice | Not Meeting Expectations | Partially Meeting Expectations | Meeting Expectations | Exceeding Expectations |
|---|---|---|---|---|
| **GMP1.2**<br>Reflect on your thinking as you solve your problem. | | | | |

# GMP1.3 Rubric

## SMP1: Make sense of problems and persevere in solving them.

| Goal for Mathematical Practice | Not Meeting Expectations | Partially Meeting Expectations | Meeting Expectations | Exceeding Expectations |
|---|---|---|---|---|
| GMP1.3<br><br>Keep trying when your problem is hard. | | | | |

Generic GMP Rubrics

A3

# GMP1.4 Rubric

## SMP1: Make sense of problems and persevere in solving them.

| Goal for Mathematical Practice | Not Meeting Expectations | Partially Meeting Expectations | Meeting Expectations | Exceeding Expectations |
|---|---|---|---|---|
| **GMP1.4** Check whether your answer makes sense. | | | | |

Generic GMP Rubrics

# GMP1.5 Rubric

**SMP1: Make sense of problems and persevere in solving them.**

| Goal for Mathematical Practice | Not Meeting Expectations | Partially Meeting Expectations | Meeting Expectations | Exceeding Expectations |
|---|---|---|---|---|
| **GMP1.5**<br><br>Solve problems in more than one way. | | | | |

Generic GMP Rubrics

# GMP1.6 Rubric

## SMP1: Make sense of problems and persevere in solving them.

| Goal for Mathematical Practice | Not Meeting Expectations | Partially Meeting Expectations | Meeting Expectations | Exceeding Expectations |
|---|---|---|---|---|
| GMP1.6 Compare the strategies you and others use. | | | | |

# GMP2.1 Rubric

## SMP2: Reason abstractly and quantitatively.

| Goal for Mathematical Practice | Not Meeting Expectations | Partially Meeting Expectations | Meeting Expectations | Exceeding Expectations |
|---|---|---|---|---|
| **GMP2.1** Create mathematical representations using numbers, words, pictures, symbols, gestures, tables, graphs, and concrete objects. | | | | |

# GMP2.2 Rubric

## SMP2: Reason abstractly and quantitatively.

| Goal for Mathematical Practice | Not Meeting Expectations | Partially Meeting Expectations | Meeting Expectations | Exceeding Expectations |
|---|---|---|---|---|
| GMP2.2 Make sense of the representations you and others use. | | | | |

# GMP2.3 Rubric

**SMP2: Reason abstractly and quantitatively.**

| Goal for Mathematical Practice | Not Meeting Expectations | Partially Meeting Expectations | Meeting Expectations | Exceeding Expectations |
|---|---|---|---|---|
| **GMP2.3** <br> Make connections between representations. | | | | |

# GMP3.1 Rubric

## SMP3: Construct viable arguments and critique the reasoning of others.

| Goal for Mathematical Practice | Not Meeting Expectations | Partially Meeting Expectations | Meeting Expectations | Exceeding Expectations |
|---|---|---|---|---|
| **GMP3.1**<br>Make mathematical conjectures and arguments. | | | | |

Generic GMP Rubrics

# GMP3.2 Rubric

## SMP3: Construct viable arguments and critique the reasoning of others.

| Goal for Mathematical Practice | Not Meeting Expectations | Partially Meeting Expectations | Meeting Expectations | Exceeding Expectations |
|---|---|---|---|---|
| **GMP3.2**<br>Make sense of others' mathematical thinking. | | | | |

Generic GMP Rubrics

# GMP4.1 Rubric

## SMP4: Model with mathematics.

| Goal for Mathematical Practice | Not Meeting Expectations | Partially Meeting Expectations | Meeting Expectations | Exceeding Expectations |
|---|---|---|---|---|
| **GMP4.1** Model real-world situations using graphs, drawings, tables, symbols, numbers, diagrams, and other representations. | | | | |

# GMP4.2 Rubric

## SMP4: Model with mathematics.

| Goal for Mathematical Practice | Not Meeting Expectations | Partially Meeting Expectations | Meeting Expectations | Exceeding Expectations |
|---|---|---|---|---|
| **GMP4.2**<br>Use mathematical models to solve problems and answer questions. | | | | |

# GMP5.1 Rubric

## SMP5: Use appropriate tools strategically.

| Goal for Mathematical Practice | Not Meeting Expectations | Partially Meeting Expectations | Meeting Expectations | Exceeding Expectations |
|---|---|---|---|---|
| GMP5.1<br>Choose appropriate tools. | | | | |

# GMP5.2 Rubric

## SMP5: Use appropriate tools strategically.

| Goal for Mathematical Practice | Not Meeting Expectations | Partially Meeting Expectations | Meeting Expectations | Exceeding Expectations |
|---|---|---|---|---|
| GMP5.2<br><br>Use tools effectively and make sense of your results. | | | | |

# GMP6.1 Rubric

## SMP6: Attend to precision.

| Goal for Mathematical Practice | Not Meeting Expectations | Partially Meeting Expectations | Meeting Expectations | Exceeding Expectations |
|---|---|---|---|---|
| **GMP6.1**<br><br>Explain your mathematical thinking clearly and precisely. | | | | |

Generic GMP Rubrics

# GMP6.2 Rubric

## SMP6: Attend to precision.

| Goal for Mathematical Practice | Not Meeting Expectations | Partially Meeting Expectations | Meeting Expectations | Exceeding Expectations |
|---|---|---|---|---|
| **GMP6.2** Use an appropriate level of precision for your problem. | | | | |

# GMP6.3 Rubric

## SMP6: Attend to precision.

| Goal for Mathematical Practice | Not Meeting Expectations | Partially Meeting Expectations | Meeting Expectations | Exceeding Expectations |
|---|---|---|---|---|
| **GMP6.3**<br>Use clear labels, units, and mathematical language. | | | | |

# GMP6.4 Rubric

## SMP6: Attend to precision.

| Goal for Mathematical Practice | Not Meeting Expectations | Partially Meeting Expectations | Meeting Expectations | Exceeding Expectations |
|---|---|---|---|---|
| **GMP6.4** Think about accuracy and efficiency when you count, measure, and calculate. | | | | |

Generic GMP Rubrics

# GMP7.1 Rubric

## SMP7: Look for and make use of structure.

| Goal for Mathematical Practice | Not Meeting Expectations | Partially Meeting Expectations | Meeting Expectations | Exceeding Expectations |
|---|---|---|---|---|
| **GMP7.1**<br><br>Look for mathematical structures such as categories, patterns, and properties. | | | | |

Generic GMP Rubrics

# GMP7.2 Rubric

**SMP7: Look for and make use of structure.**

| Goal for Mathematical Practice | Not Meeting Expectations | Partially Meeting Expectations | Meeting Expectations | Exceeding Expectations |
|---|---|---|---|---|
| **GMP7.2** Use structures to solve problems and answer questions. | | | | |

# GMP8.1 Rubric

## SMP8: Look for and express regularity in repeated reasoning.

| Goal for Mathematical Practice | Not Meeting Expectations | Partially Meeting Expectations | Meeting Expectations | Exceeding Expectations |
|---|---|---|---|---|
| **GMP8.1**<br><br>Create and justify rules, shortcuts, and generalizations. | | | | |

# Lesson 1-14  CCSS 3.MD.1, 3.MD.3

## Work Sample #1—Exceeding Expectations

This sample work meets expectations for the content standards and exceeds expectations for the mathematical practice. This child solved Problem 2 correctly, using the information in the graph and adding the minutes correctly. **3.MD.1, 3.MD.3** For Problem 1, the child gave six true statements based on the graph, including comparison statements (statements 5 and 6). For Problem 2, the explanation shows how to use the differences in the starting times and the information in the graph to find that Cheryl arrived at school first, noting the 5-minute difference in the arrival times for Cheryl and Carlos. **GMP4.2**

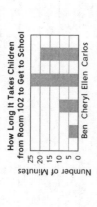

**How Long It Takes Children from Room 102 to Get to School**

1. Carefully look at the title, labels, and bars on the graph. Write at least 5 things you know from the graph.

① Ben takes 5 min.
② Cheryl takes 10 min.
③ Carlos takes 20min.
④ Ellen takes 25 min.
⑤ Ellen takes the most time.
⑥ Ben takes the least time.

2. Carlos leaves for school at 8:00. Cheryl leaves five minutes later.

a. Who gets to school first? Cheryl

b. Explain how you figured it out.
Well Carlos leves at 8:00.
And cheryl leaves at 8:05.
So cheryl getstore first
Cadded 5 mins to cheryl time.
It was 15 mins. That is 5 mins
less than carlos

Evaluated Children's Work Samples    **A23**

**NOTE** The wording and formatting of the problem on the sample work may vary slightly from that of the actual problem your children will complete. These minor discrepancies do not affect the implementation of the problem.

# Lesson 1-14   CCSS 3.MD.1, 3.MD.3

## Work Sample #2—Partially Meeting Expectations

This sample work does not meet expectations for the content standards and partially meets expectations for the mathematical practice. This child identifies Cheryl as the first to arrive at school, stating that he or she used the graph but giving no indication of the information that was used or how to add minutes. **3.MD1, 3.MD.3** For Problem 1, the child gave only three accurate statements (statements 3–5) using information from the graph. For Problem 2, the child states that the graph was used to determine that Cheryl arrives first, but there is no explanation of the child's thinking. **GMP4.2**

**NOTE** The rubric for GMP4.2 appears on page 109 of the *Teacher's Lesson Guide*.

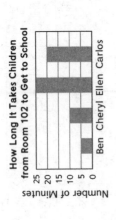

**How Long It Takes Children from Room 102 to Get to School**

Number of Minutes

Ben   Cheryl   Ellen   Carlos

1. Carefully look at the title, labels, and bars on the graph. Write at least 5 things you know from the graph.

1. ben is the first one to get to school

2. ellen is the last one to get to school

3. I takes no one less then five minutes to get to song

4. I takes no one more than 35 minutes to get to school

5. I takes cheryl 10 minutes to get to school

2. Carlos leaves for school at 8:00.   Cheryl leaves five minutes later.

   a. Who gets to school first?   Cheryl

   b. Explain how you figured it out.

   I used the graph.

Evaluated Children's Work Samples

**A24**

# Lesson 1-14  CCSS 3.MD.1, 3.MD.3

## Work Sample #3—Meeting Expectations

This sample work does not meet expectations for the content standards but meets expectations for the mathematical practice. Although Cheryl is named as the first to arrive at school, the work indicates a misconception (8:00 + 5 min = 13 minutes) and includes a computational error in adding minutes (8:05 + 10 = 8:10). **3.MD.1, 3.MD.3** For Problem 1, the child gave five correct statements using information from the graph. Although this work includes a computational error in Problem 2, the explanation shows how the child used the differences in the starting times and the information in the graph to determine that Cheryl arrives first. **GMP4.2**

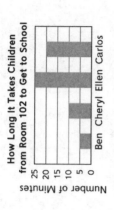

**How Long It Takes Children from Room 102 to Get to School**

Ben   Cheryl   Ellen   Carlos

Number of Minutes — 0, 5, 10, 15, 20, 25

1. Carefully look at the title, labels, and bars on the graph. Write at least 5 things you know from the graph.

   I know that it takes 5 minutes for Ben to get to school.
   It takes Cheryl 10 minutes.it takes Ellen 25 minutes.
   It takes Carlos 20 minutes.
   The bars are for how many minutes it takes for the children to get to school.

2. Carlos leaves for school at 8:00. Cheryl leaves five minutes later.

   a. Who gets to school first? _Cheryl_

   b. Explain how you figured it out.

   What I did was added 8:00+5=8:05 then
   added 8:05+10=8:10 for Cheryl and then added
   8:00+20=8:20 for Carlos,so Cheryl won.

   8:00
   5 min

   13 minutes

**NOTE** The wording and formatting of the problem on the sample work may vary slightly from that of the actual problem your children will complete. These minor discrepancies do not affect the implementation of the problem.

# Lesson 3-14 CCSS 3.NBT.2

## Work Sample #1—Partially Meeting Expectations

This sample work does not meet expectations for the content standard but partially meets expectations for the mathematical practice. The work provides no evidence that the child knows how to correctly use expand-and-trade subtraction. **3.NBT.2** Although the child recognizes that Mia's mistake occurred in the hundreds place, there is no further explanation that it is necessary to trade a hundred from the hundreds place and add 10 tens to the tens place. **GMP3.2**

$$
\begin{array}{r}
400 \quad \overset{140}{40} \quad 12 \\
552 \rightarrow 500 + 50 + 2 \\
-153 \rightarrow 100 + 50 + 3 \\
\hline
400 + 90 + 9 = 499
\end{array}
$$

"Oops," said Mia, "I didn't cross out 500 and write 400."
Explain **why** not changing 500 to 400 is a mistake.

(Hint: Use what you know about place value in your answer.)

It was a Mistake because she had to cross out a hundred so her sum is not right ether

## Work Sample #2—Not Meeting Expectations

This sample work meets expectations for the content standard but does not meet expectations for the mathematical practice. The child used expand-and-trade subtraction to find the correct answer. **3.NBT.2** Although the child recognizes that Mia's failure to "cross that out would goof up the whole problem," there is no mention of the hundreds or tens places or any reference to place-value concepts in the response. **GMP3.2**

$$
\begin{array}{r}
\overset{140}{40} \quad 12 \\
552 \rightarrow 500 + 50 + 2 \\
-153 \rightarrow 100 + 50 + 3 \\
\hline
400 + 90 + 9 = 499
\end{array}
$$

"Oops," said Mia, "I didn't cross out 500 and write 400."
Explain **why** not changing 500 to 400 is a mistake.

(Hint: Use what you know about place value in your answer.)

She forgot to cross out the 500 and make it a 400 And if you don't cross that out that would goof up the whole problem. And that's Mia's mistake!

$$
\begin{array}{r}
\overset{140}{\cancel{4}}\overset{4}{\cancel{5}}0 \quad 2 \\
552 = 500 + 50 + 2 \\
-153 = 100 + 50 + 3 \\
\hline
399 \quad 300 + 90 + 9
\end{array}
$$

# Lesson 3-14   CCSS 3.NBT.2

## Work Sample #3—Meeting Expectations

This work meets expectations for the content standard and for the mathematical practice. The child's explanation of Mia's mistake provides a detailed description of how to use expand-and-trade subtraction.

**3.NBT.2** The work meets expectations for the mathematical practice because the child makes appropriate use of place-value concepts in making sense of Mia's mistake (i.e., "the hundreds were not regrouped"). The child also explains that "you could not subtract 40 − 50 so she [Mia] took 1 hundred for the tens." **GMP3.2**

$$
\begin{array}{r}
140 \\
40 \quad 12 \\
552 \rightarrow \quad 500 + 50 + 2 \\
-153 \rightarrow \quad 100 + 50 + 3 \\
\hline
400 + 90 + 9 = 499
\end{array}
$$

"Oops," said Mia, "I didn't cross out 500 and write 400."
Explain **why** not changing 500 to 400 is a mistake.

(Hint: Use what you know about place value in your answer.)

She needed to make 500 to 400 because for the tens you could not subtract 40-50 So she took 1 hundred for the tens and she fogot to cross out the 500 and make it 400. It is a mistake because the hundreds where not regrouped and if she did regroup the hundreds wich she will probly do her estment will be a good one.

**NOTE** The wording and formatting of the problem on the sample work may vary slightly from that of the actual problem your children will complete. These minor discrepancies do not affect the implementations of the problem.

# Lesson 5-12 CCSS 3.OA.5

## Work Sample #1—Meeting Expectations

This sample work meets expectations for the content standard and for the mathematical practice. This child's labeled array provides evidence that the child understands that $6 \times 7 = 2 \times 7 + 2 \times 7 + 2 \times 7 = 14 + 14 + 14 = 42$. Furthermore, the written explanation implies that he or she understands that $14 + 14 = 28$ and $28 + 14 = 42$. **3.OA.5** This child used a helper fact, $4 \times 7 = 28$, to develop an efficient multiplication strategy for $6 \times 7$. The child provides a clear drawing and explanation to support use of a different strategy. "I did $4 \times 7$ and then added 2 more groups of 7 and I got $6 \times 7 = 42$." **GMP6.4**

2. Choose at least one other efficient multiplication strategy, such as doubling or near squares, to solve **6 x 7 = ?**. Use (pictures) and (words) to show how you solved the problem.

(Hint: What helper fact can you use?) Helper fact: $4 \times 7 = 28$

I did 4x7 and then
added 2 more
groups 7 and I got

$6 \times 7 = 42$

X XXXXXX→14
X XXXXXX→14
X XXXXX→14
X XXXXX X
XX XXXX→14
XXXXXX→28
$42$

$6 \times 7 = 42$

---

**NOTE** The rubric for GMP6.4 appears on page 523 of the *Teacher's Lesson Guide*.

## Work Sample #2—Partially Meeting Expectations

This sample work meets expectations for the content standard and partially meets expectations for the mathematical practice. This child indicates with words and numbers that $6 \times 7 = 4 \times 7 + 14 = 28 + 14 = 42$. **3.OA.5** While this child used a helper fact, $4 \times 7 = 28$, to develop an efficient multiplication strategy for $6 \times 7$, the drawing and explanation of the strategy are incomplete. The array shows 6 groups of 7 but does not show how the groups are decomposed into 28 and 14. The explanation uses "one more group of 14 to get 42" but fails to clearly show how this strategy results in 6 groups of 7 [$6 \times 7$]. **GMP6.4**

2. Choose at least one other efficient multiplication strategy, such as doubling or near squares, to solve **6 x 7 = ?**. Use (pictures) and (words) to show how you solved the problem.

(Hint: What helper fact can you use?)

I know $4 \times 7 = 1528$ and
I add One more group of
14 to get 42.

X XXX XX
XX X XXX
XX XX XXX
X XX XXX
XXX XXX
XXX XX XXX

**NOTE** The rubric for GMP6.4 appears on page 523 of the *Teacher's Lesson Guide*.

# Lesson 5-12 CCSS 3.OA.5

## Work Sample #3—Not Meeting Expectations

This work does not meet expectations for the content standard or for the mathematical practice. Although this child drew an array that showed 6 rows of 7, the child did not show how to decompose or combine groups to solve the problem. **3.OA.5** The child did not use a different strategy than the one in Problem 1 to solve 6 × 7. **GMP6.4**

**NOTE** The wording and formatting of the problem on the sample work may vary slightly from that of the actual problem your children will complete. These minor discrepancies do not affect the implementation of the problem.

2. Choose at least one other efficient multiplication strategy, such as doubling or near squares, to solve 6 × 7 = ?. Use pictures and words to show how you solved the problem.

(Hint: What helper fact can you use?) Helper fact: 5X7=35

I used the helper fact 5x7. I know that 5X7 is 35, so 6X7=42.

X X X X X X X
X X X X X X X
X X X X X X X
X X X X X X X
X X X X X X X
X

**NOTE** The wording and formatting of the problem on the sample work may vary slightly from that of the actual problem your children will complete. These minor discrepancies do not affect the implementation of the problem.

# Lesson 7-13   CCSS 3.NF.3, 3.NF.3d

## Work Sample #1–Partially Meeting Expectations

This sample work does not meet expectations for the content standards but partially meets expectations for the mathematical practice. This work does not show an understanding that the size of the whole of each fraction needs to be considered when comparing fractions. **3.NF.3, 3.NF.3d** This child argues that both Demtrius and Emma could be right, but neither drawings nor words are used to make clear arguments. The drawing for Demitrius's statement does not show a pizza that is larger than Emma's because only one pizza is drawn. The written arguments do not claim that Demitrius could be correct if his pizza had been bigger than Emma's, or that Emma could be correct if they had started with the same size pizza. **GMP3.1**

**NOTE** The rubric for GMP3.1 appears on page 717 of the *Teacher's Lesson Guide.*

Use words and pictures to show that Demitrius could be right.

Demitrius could be right because he could have calculated what halve of the pizza could be.

Use words and pictures to show that Emma could be right.

Emma could be right because they both ate 1/2 of a pizza

**NOTE** The wording and formatting of the problem on the sample work may vary slightly from that of the actual problem your children will complete. These minor discrepancies do not affect the implementation of the problem.

# Lesson 7-13   CCSS 3.NF.3, 3.NF.3d

## Work Sample #2—Partially Meeting Expectations

This sample work does not meet expectations for the content standards but partially meets expectations for the mathematical practice. While part of this child's statement showed that he or she recognized that a comparison of fractions depends on the size of the whole for each fraction, the statement that Demetrius could have had more pieces indicates a lack of understanding that the size of the pieces depends on the size of the whole, and that more smaller-size pieces does not necessarily mean a larger amount. **3.NF.3, 3.NF.3d** The pictures show how each claim could be correct. This child also included written arguments. The child correctly argued that "Demitrius can be right because he could have had a bigger pizza" and "Emma could be right because they might have the same size pizza." However, the child also wrote that, "[Demitrius] could have more pieces" and wrote about the "amount of slices" in the argument for Emma's claim. Since eating more pieces only results in eating more pizza if the size of the pieces is considered along with the size of the whole, this argument and the similar argument about the "amount of slices" are not clear. Had the "amount of slices" arguments not been included, this work would have met expectations. **GMP3.1**

**NOTE** The rubric for GMP3.1 appears on page 717 of the *Teacher's Lesson Guide*.

Use words and pictures to show that Demitrius could be right.

Demitrius can be right because he could have had a bigger pizza, or he could have more pieces.

Use words and pictures to show that Emma could be right.

Emma could be right because They might have ordered the same size pizza, or they can the same amount of slices

NOTE The rubric for GMP3.1 appears on page 717 of the *Teacher's Lesson Guide*.

NOTE The wording and formatting of the problem on the sample work may vary slightly from that of the actual problem your children will complete. These minor discrepancies do not affect the implementation of the problem.

# Lesson 7-13 CCSS 3.NF.3, 3.NF.3d

## Work Sample #3—Exceeding Expectations

This sample work meets expectations for the content standards and exceeds expectations for the mathematical practice. This child recognized that a comparison of fractions depends on the size of the whole for each fraction. **3.NF.3, 3.NF.3d** Because there are clear arguments for both claims with words and with pictures, this work exceeds expectations. GMP3.1

**Use words and pictures to show that Demitrius could be right.**

Demitrius could be Right IF
His Pizza was Larger than
emma's.

emma        Demitrius

**Use words and pictures to show that Emma could be right.**

EMMA Cold be Right IF
There Pizza's were the
Same.

emma  Demi.

Evaluated Children's Work Samples     **A32**

**NOTE** The rubric for GMP7.1 appears on page 858 of the *Teacher's Lesson Guide*.

**NOTE** The wording and formatting of the problem on the sample work may vary slightly from that of the actual problem your children will complete. These minor discrepancies do not affect the implementation of the problem.

# Lesson 9-8 CCSS 3.OA.9

## Work Sample #1—Partially Meeting Expectations

This sample work does not meet expectations for the content standard but partially meets expectations for the mathematical practice. In Problem 2 the number sentences show the results of doubling a factor, but Problem 1 is not complete and the explanation in Problem 2 that "the first product is the next ones factor" does not describe the pattern.

**3.OA.9** While this child provided examples of how a tripled factor results in a tripled product ($2 \times 6 = 12$ and $6 \times 6 = 36$), the explanation of the pattern is not complete It does not describe a connection between tripling a factor and tripling a product, and the statement that "they add a number to the factor and they get the answer" is incorrect. **GMP7.1**

1. Explore what happens to the product when you double a factor. For example, multiply 4 x 5. Then double the 4 and multiply 8 x 5; then double the 5 and multiply 4 x 10.

   $$10 \times 4 = 40$$

2. Try doubling one of the factors in other multiplication facts. Show your work.
   Describe a pattern you see when you double factors.

   the first product is the next ones factor.

   $$6 \times 2 = 12$$
   $$12 \times 2 = 24$$
   $$24 \times 2 = 48$$

3. Based on your work with doubling factors, predict what will happen when you triple a factor. Explain how you would convince someone that your prediction is true for all multiplication facts.

   $$2 \times 6 = 12$$
   $$6 \times 6 = 36$$
   $$18 \times 6 = 108$$

   they add a number to the factor and they get the answer.

NOTE The rubric for GMP7.1 appears on page 858 of the *Teacher's Lesson Guide.*

NOTE The wording and formatting of the problem on the sample work may vary slightly from that of the actual problem your children will complete. These minor discrepancies do not affect the implementation of the problem.

# Lesson 9-8    CCSS 3.OA.9

## Work Sample #2—Meeting Expectations

This sample work meets expectations for the content standard and for the mathematical practice. This child gives number sentences that show the results of doubling a factor in Problems 1 and 2 and accurately describes the doubling pattern in Problem 2 by saying, "This answer is doubled too." **3.OA.9** While the explanation of this child's prediction ("you multiple 3 times and so the number will get 3 times bigger") does not use the term *multiple* correctly or use the terms *factor* or *product* at all, the explanation does indicate that when a factor is tripled the product will be tripled. The child also provides two examples that justify the prediction.

GMP7.1

1. Explore what happens to the product when you double a factor. For example, multiply 4 × 5. Then double the 4 and multiply 8 × 5; then double the 5 and multiply 4 × 10.

$$4 \times 5 = 20$$
$$8 \times 5 = 40$$
$$4 \times 10 = 40$$

2. Try doubling one of the factors in other multiplication facts. Show your work.
Describe a pattern you see when you double factors.

$$6 \times 5 = 30 \quad \text{It's doubled}$$
$$12 \times 5 = 60.$$
$$4 \times 3 = 12 \quad \text{This answer's doubled}$$
$$8 \times 3 = 24. \text{ too.}$$

3. Based on your work with doubling factors, predict what will happen when you triple a factor. Explain how you would convince someone that your prediction is true for all multiplication facts.

$$6 \times 5 = 30$$
$$18 \times 5 = 90$$

$$\begin{array}{r} 4 \\ 18 \\ \times 5 \\ \hline 90 \end{array}$$

$$4 \times 3 = 12$$
$$12 \times 3 = 36$$

You multiple 3 times and So the number will get 3 times bigger.

Evaluated Children's Work Samples    A34

NOTE The wording and formatting of the problem on the sample work may vary slightly from that of the actual problem your children will complete. These minor discrepancies do not affect the implementation of the problem.

# Lesson 9-8 CCSS 3.OA.9

## Work Sample #3—Exceeding Expectations

This sample work meets expectations for the content standard and exceeds expectations for the mathematical practice. This child gives number sentences that show the results of doubling a factor in Problems 1 and 2, and the doubling pattern is accurately described in Problem 2. **3.OA.9** The child predicted that the pattern found in doubling a factor can be extended to tripling a factor and provides examples that support this prediction. Furthermore, the child provides reasoning for why the prediction works by saying, "when you triple a factor you just add the product three times." **GMP7.1**

NOTE The rubric for GMP7.1 appears on page 858 of the *Teacher's Lesson Guide*.

**1.** Explore what happens to the product when you double a factor.
For example, multiply 4 × 5. Then double the 4 and multiply 8 × 5; then double the 5 and multiply 4 × 10. Show your work.

When you do 4×5=20 and double the 4 and do 8×5=40. The answer doubled too.

**2.** Try doubling one of the factors in other multiplication facts. Show your work.
Describe a pattern you see when you double factors.

9×6=54 and when I double the 9 I get 18×6=108 and the answer doubles too.

**3.** Based on your work with doubling factors, predict what will happen when you triple a factor. Explain how you would convince someone that your <u>prediction</u> is true for all multiplication facts.

I think when you triple a factor, you just add the product three times. Say for instance the factors were 7×6=42 and you triple the 7-into a 21 and do 21×6=126 if you knew 42×3=126 you would have got the same thing